THE FUTURE OF
4D PRINTING

www.royalcollins.com

THE FUTURE OF
4D PRINTING
INNOVATIONS AND APPLICATIONS

KEVIN CHEN

Books Beyond Boundaries

ROYAL COLLINS

The Future of 4D Printing: Innovations and Applications

KEVIN CHEN

First published in 2024 by Royal Collins Publishing Group Inc.
Groupe Publication Royal Collins Inc.
550-555 boul. René-Lévesque O Montréal (Québec) H2Z1B1 Canada

10   9   8   7   6   5   4   3   2   1

ISBN: 978-1-4878-1184-6

To find out more about our publications, please visit www.royalcollins.com.

CONTENTS

CONTENTS

CONTENTS

CONTENTS

FOREWORD

Compared to the prevailing 3D printing, the concept of 4D printing, which has been around for more than two years, seems a bit deserted. What is 4D? Where did the additional "D" come from?

Before reading the following text to explore 4D printing, let me tell you a short story.

One day, a man named Adam and a woman named Eve stole the forbidden fruit of the tree of knowledge, of good and evil, in the Garden of Eden. Although they didn't understand design, together they created a work which I think is the greatest in this world—the human!

In the entire process, the sperm of a man and the egg of a woman matched and combined to form a brand-new cell, and the inside raw materials that make up the human were the 23 pairs of chromosomes—46 chromosomes in total. Each chromosome carries a certain number of design factors that we call "genes." Genes support the basic structure and performance of human life and store all the information about race, blood type, pregnancy, growth, and the apoptosis process of each person.

After the materials were fully prepared, the first person was "printed," and the carrier of this printing was the mother of all humans—Eve. Since then, with the extension of the time dimension and "catalysis" of nature, humans have begun to change through birth, growth, disease, aging, and death. They also started to further deduce the reproduction of lives and the important physiological processes, such as cell division and protein synthesis. All these changes and processes repeating in cycles may not be new to us, and we know them well, for sure.

So, now, that we are finished with the storytelling, let's get back to business. What is 4D printing?

The above short story implies a common understanding of 4D printing, that is, the genetic code contained in each of our chromosomes is the most primitive design programming code of all of us, humans. The growing process of our life is the organizational change generated by humans, the 4D printed object, based on the fourth dimension of time and the "catalysis" of nature.

In other words, the biggest difference between 4D printing and 3D printing is that 3D printing, like all manufacturing processes nowadays, is the end of creation. However, 4D printing is the beginning of creation, just like a human birth.

Of course, this is just a small analogy to 4D printing based on personal understanding. To unveil it and give you a full and more comprehensive picture, let's take a look at how the versatile Baidu* explains this science fiction phenomenon. According to the information from Baidu encyclopedia entries, 4D printing has one more "D" than 3D printing, which is the dimension of time. People can set the model and time through software, and the transformable material will self-transform into the desired shape over the specified time. To be precise, 4D printing is a kind of material that can transform on its own, in which the design is directly embedded. Without connecting any complicated electromechanical equipment, it can self-fold into the corresponding shape, in line with the product design.

Is this really how 4D printing works? Let's take a look at its past and present.

---

* Baidu is a popular online search engine in China.

FOREWORD

On February 26, 2013, at the TED2013 conference in Long Beach, California, Skylar Tibbits, a lecturer from the Department of Architecture, Massachusetts Institute of Technology (MIT), used a novel combination of materials on a 3D printer to create a string of strands. When the strand was placed in the water, its shape changed and formed the letters "MIT." This is a technology that produces the effect by combining water-absorbing polymeric materials with alkaline plastics and is defined as 4D printing. "What we're saying here is, you design something, you print it, it evolves. It's like naturally embedding smartness into the materials," Tibbits said in an interview.

On October 8, 2014, the US *Foreign Affairs* bimonthly magazine published an article entitled Preparing for the 4D Printing Revolution. The author, Nayef Al-Rodhan, an honorary fellow at St. Antony's College, Oxford University, said in the article that the possible applications of 4D printing were endless, and the real future of digital manufacturing lay in the fourth dimension, which was to print objects in programmable materials that could self-transform over time.

Based on this, we may soon see that science fiction could turn into reality.

Imagine that the underground of every city we live in is full of piping systems; the fixed capacity and high maintenance costs have always been troubling the constructors of every city, too. However, with the arrival of 4D printing, by using programmable materials, each pipeline will be able to adapt to the changing environment, and its capacity and flow rate can be adjusted through expanding or contracting motion. It can even self-repair when damaged or self-decompose when scrapped.

What is more amazing is that the application of 4D printing in the field of biomedicine will greatly change existing medical conditions. In the era of 3D printing, we have been able to use 3D printed hearts, limb skeletons, and other human organs to replace the aging and diseased organs of the human body and keep humans healthy. The emergence of 4D printing will undoubtedly bring us more possible optimized solutions. For example, with the change of a heart stent, no surgery is needed if we can achieve a 4D printed liquid stent and inject it into the programmed smart material via the blood circulation system. When it reaches the designated part of the heart, it will self-assemble into a new heart stent.

Moving forward, we could even use 4D printing to gradually replace our aging organs, thereby slowing down our aging process and the progression of diseases. As 4D printing technology emerges, the potential optimization solutions in the medical field will become more diverse and wondrous.

Now, let us formally explore this 4D printing technology filled with infinite possibilities and the era of 4D printing we are about to enter.

1

THE CONCEPT

Over a thousand years ago, the advent of 2D movable type printing allowed us to "copy" text on paper, which essentially is what today's 2D printing represents—a flat duplication.

Subsequently, the emergence of 3D printing technology enabled us to print objects in three dimensions. For example, with a 3D printer, we can produce a three-dimensional flower.

So, is there a technology that allows this flower to keep pace with real-time, blossoming gradually with the sunrise, just like a real flower?

Indeed, there is—4D printing.

SECTION 1 | 4D PRINTING IS HERE

The first introduction of 4D printing technology occurred on February 25, 2013, at the TED 2013 conference in California, US. It was presented by Skylar Tibbits, a scientist from the Self-Assembly Lab at MIT, through a demonstration of products self-transforming and assembling, officially bringing 4D printing technology to our attention and establishing an understanding of this new technology.

In layman's terms, 4D printing adds a time dimension to 3D printing—by employing programmable materials to control the transformable elements within a 3D printed object, the shape and properties of the object can change over time in response to environmental stimuli such as light, heat, sound, and magnetism, enabling automatic transformation, self-repair, and self-assembly.

In the traditional concept of dimensions, it is difficult for us to cognize and comprehend the concept of the fourth dimension. From 1-dimension to 2-dimension to 3-dimension, these are the dimensions based on physical space and can be intuitively seen with our sensory vision (Figure 1-1). However, the arrival of the fourth dimension seems to transcend our current understanding of three-dimensional space. Its newly added dimension, time, is what we have never seen or touched but a concept that always surrounds us. This is one of the explanations of 4D printing now—4D printing is printing time.

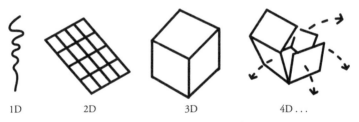

| 1D | 2D | 3D | 4D . . . |

Figure 1-1 From 1-dimension to 4-dimension

This is not merely a speculative concept from futurology but a new technology researched by MIT, demonstrated by Skylar Tibbits at the TED conference. Throughout the demonstration, a 4D printed strand in multi-material completed self-transformation and self-assembly on its own while being dipped in water. As introduced at the conference, the printer used to print this composite material is not a new or magical technology; it is a 3D printer. However, the raw materials used are truly remarkable, consisting of a piece of plastic and a layer of "smart" material capable of absorbing water. Tibbits thinks that "the printing process is not something new, and the core value is the change that occurred after printing." That is exactly the key to 4D printing.

Skylar Tibbits, the director of the 4D printing project, is only 28 years old and is an architect and a computer scientist. The 4D printing method he invented can only work in the water at this time, which means that the triggering medium is limited to water only. According to Tibbits, self-assembly is the core behind 4D printing, and it has been used on the nanoscale for many years, just like the self-folding proteins that are now widely applied (Figure 1-2). If the applications of the technology are further extended to the construction of bridges, tall buildings, biological organs, and daily necessities, this would turn our perceptions upside down. The extension implies that the applications of 4D printing are not far away from our daily lives, but they are close enough to touch, and soon, they will comprehensively reach us.

Protein folding technology is currently a very important technique in the field of biology. Gregory Weiss, a biochemist at the University of California, stated in his article "Edible Science: Ten Things You Didn't Know about Food" that they could "rejuvenate" cooked eggs using protein folding technology. However, studying protein folding is not just about rejuvenating eggs; its more important purpose is to reveal the second set of genetic codes within living organisms—the folding code. Proteins within the human body and other organisms are composed of various amino acids. As one of the most important functional carriers of life, proteins play a key role in numerous biological activities. However, proteins often need to fold into specific three-dimensional structures to perform their functions—amino acids arranged in a long chain, when placed in water, will fold into a stable three-dimensional structure within a second. This is the self-folding nature of proteins.

The concept of 4D technology is inspired by this self-folding technique of proteins, drawing inspiration from biology. Under the guidance of Professor Arthur Olson, a molecular biology professor, Tibbits demonstrated 4D printing technology by designing a set of self-assembling parts using 3D printing technology combined with embedded magnets. Simply shaking the beaker forcefully, the parts in the beaker would assemble themselves into a 3D model of the poliovirus as if they were conscious. This experiment is the prototype of 4D printing technology and serves as proof of the feasibility of 4D printing.

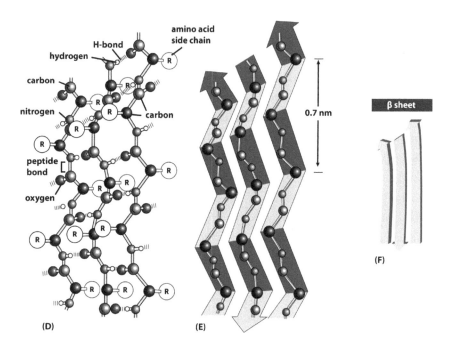

Figure 1-2 Self-folding proteins

From that point on, Tibbits began to explore the commercial applications of 4D printing and visited the project manager of the world's largest 3D printer manufacturer—Stratasys. In their communication, he got to know that the company's research in the field of printing materials had made a breakthrough. A macromolecule polymer that could self-transform in the water was invented. The unique feature of the new material is that it can self-extend to twice its length when exposed to water.

It is this new material that escalated Tibbits' research and development of the new technology, which is 4D printing. Along with the design software and

3D printer, he mixed the two materials, extensible and non-extensible, and printed them into a strand, then immersed them in the water. Immediately, the strand underwent a fascinating process of self-bending and self-transforming on its own and formed the abbreviation of MIT (Figure 1-3) as if it had a self-thinking ability.

Figure 1-3 4D printed "MIT"

On this basis, Tibbits conducted the second experiment, which was more complex and had more complicated algorithms and designs. If the first exploring experiment was on the plane level, this time, it was at the level of three-dimensional space, that is, the self-assembly and self-transformation of objects in three-dimensional space (Figure 1-4). The second experiment was also about printing the transformable design into a strand and placing it in the water. With the help of water as the triggering medium, it managed to self-assemble and self-transform into a cube.

To Tibbits, significant success and breakthroughs have been accomplished in the second experiment, which marked the arrival of 4D printing. This technology will soon walk out of the lab and come straight into our daily lives. The changes brought by 4D printing will be more profound and thorough. This will include the troubling underground drainage system in the cities. If the technology is applied to build contractible drainage pipelines, the pipelines will automatically become larger to better facilitate drainage when a typhoon or a rainstorm comes. Then, the pipelines will automatically contract to their original size (Figure 1-5) when the water flow slows down. This implies that the urban infrastructure built on 4D printing technology will create a true sponge city.

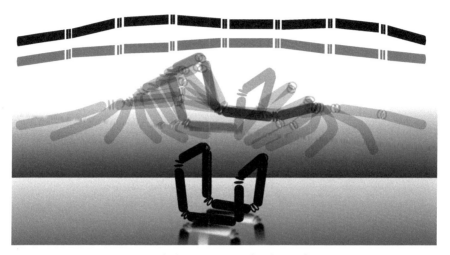

Figure 1-4 4D printing technology of MIT

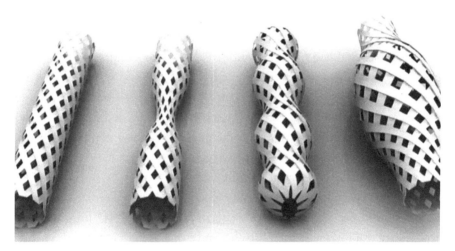

Figure 1-5 Self-transformable pipelines

The concept of a sponge city was proposed by Chinese experts as a new concept for urban construction and governance, mainly aimed at rain and flood management. It means that a city can have good resilience like a sponge in adapting to environmental changes and responding to natural disasters caused by rainfall, which can also be understood as a "water-resilient city." It enables the city to have good "elasticity" in adapting to environmental changes

6

and dealing with natural disasters. If expressed in international common terminology, it is "low impact development for stormwater management," which involves absorbing, storing, infiltrating, and purifying water during rainfall and, when necessary, releasing and utilizing the stored water, allowing for the free migration of rainwater in the city.

Such a new concept of urban construction imposes higher demands on the drainage system. If utilizing self-transforming pipes constructed with 4D printing technology, these pipes can also bend, twist, and transform at specific locations according to the underground space conditions. We don't have to worry about them bursting underground or being troubled by the discharge capacity during peak flood times; they will automatically transform according to the volume of water flow. Moreover, in regions prone to geological disasters, these pipes can even self-assemble and repair. It can be said that 4D printing is not only a new technology but also a revolutionary new material technology for human society.

SECTION 2 | FROM 3D PRINTING TO 4D PRINTING

Compared to 4D printing, 3D printing seems to be a more familiar concept to us.

3D printing, as the name suggests, is three-dimensional printing. Whether it's two-dimensional flat printing or three-dimensional solid printing, both are essentially printing technologies. The difference is that flat printing produces documents in a flat shape, conveying information without possessing any practical functionality. In contrast, unlike flat-printed documents, 3D printing can directly create functional objects.

3D printing requires the three-dimensional shape information of the item to be printed to be input into a file that the 3D printer can interpret. Once the 3D printer reads the file, it prints the solid shape by accumulating material layer by layer. It can be said that the three-dimensional shape is the basis of its function; by printing the shape, the functionality is also created.

Moreover, as opposed to "subtractive manufacturing," 3D printing is also known as "additive manufacturing." In the current stage of manufacturing, the commonly used material processing technology is mostly "subtractive manufacturing," which involves removing, cutting, and assembling raw materials to give them specific shapes and functionalities. "Additive manufacturing," on the other hand, builds up the raw material layer by layer into a specific shape to achieve a specific function.

The additive manufacturing process primarily includes two stages: three-dimensional design and layer-by-layer printing. First, a model is created using computer modeling software; then the completed 3D model is divided into layered sections, guiding the printer to print layer by layer. Compared to traditional subtractive manufacturing methods, additive manufacturing undoubtedly has many advantages.

To shorten the time of production and manufacturing, improving efficiency. Using traditional methods to manufacture a model typically takes several days, depending on the size and complexity of the model, whereas 3D printing technology can reduce this time to just a few hours. Therefore, compared to subtractive manufacturing, additive manufacturing is particularly suitable for making complex-shaped parts. Of course, this is also affected by the performance of the printer and the size and complexity of the model.

To increase the efficiency of raw material use. Compared to traditional metal manufacturing techniques, 3D printing produces fewer by-products when making metal parts. With the advancement of printing materials, "net shape" manufacturing could become a more environmentally friendly processing method.

To achieve complex structures to enhance product performance. Traditional subtractive manufacturing has limitations in processing complex shapes and internal cavity structures, whereas 3D printing can improve product performance by manufacturing complex structures, offering unmatched advantages in aerospace, mold processing, and other fields.

For example, a 3D printer can produce many shapes, creating different parts each time like a craftsman. For traditional machine tool production lines, processing parts of different shapes requires complex adjustments to the production line. Therefore, additive manufacturing is especially suitable for customized, non-mass-produced items.

Initially, 3D printing was mainly used in mold manufacturing and industrial design to create models, and later it gradually started being used for the direct production of certain products or parts, including applications in vertical fields like aerospace, engineering construction, medical, education, geographic information systems, automotive, etc.

In 2015, National Aeronautics and Space Administration (NASA), based on 3D printing technology, printed the head of an aviation rocket engine. This significantly reduced the number of parts and welds, thereby lowering the probability of rocket engine failure, shortening the iteration cycle, and reducing costs.

In Dubai, the government chose to use 3D printing to construct government buildings. The main work of 3D printed architecture is done by machines, with integrated formation, fast construction speed, and the workers mainly operating and inspecting the 3D printers, thus requiring less manpower than traditional construction.

In 2019, Tel Aviv University in Israel announced that its laboratory had 3D printed a "heart," not just an exterior model but the world's first 3D printed, vascularized engineered heart using the patient's own cells and biological materials, i.e., a three-dimensional artificial heart with vascular tissue.

In May 2020, the successful maiden flight of China's Long March 5B carrier rocket not only carried China's new generation of manned spacecraft test ships but also a "3D printer." This marked China's first space 3D printing experiment and the world's first in-space continuous fiber-reinforced composite material 3D printing experiment.

With the change of the entire society and business forms brought by the mobile Internet, including the rise of Makers,* they have contributed to the explosion of the 3D printing industry, most notably so in 2013.

---

* The term refers to a group of people who love technology and its practice and enjoy sharing technology and exchanging ideas. The community (Makerspace) with Makers as the main body has become the carrier of Makers' culture.

Of course, the outbreak of 3D printing is the trend of the industry's evolution, and it being able to be recognized and accepted quickly by the public is related to our "dreams." Most of us might have heard of the fairy tale *The Magical Paintbrush—Story of Ma Liang*[†] when we were kids, and we yearned to have such a magic paintbrush so that we could draw our dreams and make them come true. Although this is just a fairy tale, when combined with 3D printing, we seem to see the moment that it could turn into reality. We can print any item that we want via 3D printing, no matter if it is a human, organ, food, or weapon; as long as we can design a three-dimensional model on a computer, then a 3D printer will print it out for us.

It doesn't matter if we don't know how to use a three-dimensional design on a computer. We only need to scan the entity of the item that we want to print with a 3D scanner, and a 3D model will be automatically generated through the computer system and then sent to a 3D printer for printing.

However, the advancement of technology sometimes evolves at a pace beyond our understanding. While we are still familiarizing ourselves with the concept of professional 3D printers, 3D printing technology has quietly developed to the desktop level, especially after the integration with artificial intelligence (AI), particularly with AI software that comes with design capabilities. Regardless of whether we possess professional design and operational knowledge related to 3D printing software, we can leverage AI software to help us realize design ideas, and this type of customized 3D printing is evolving toward becoming accessible in every household. Not only that, but smaller 3D printers have also emerged before our eyes. They are truly like the magical brush in Ma Liang's hand; we only need to hold a 3D printing pen and draw directly. The drawing process is simultaneously a printing process.

Therefore, we can have a good understanding that 3D printing, after integrating AI, is like Ma Liang's magic paintbrush, which can not only print models

---

[†] This is a fairy tale from China. The story is about a little boy, Ma Liang, who obtains a magic paintbrush by coincidence. No matter what he draws using the paintbrush, it becomes real. The little boy uses this magic paintbrush and his wisdom to help the poor against the oppression of the emperor.

but also ideas for us. With 3D printing, we can design a special cup based on our ideas and creativity and turn it into reality. We can also create a unique pair of shoes and privately customize it. We can even print a table of edible and artistic-shaped food. As we can see, the development of 3D printing has brought great influence, changes, and imagination to our future life and business forms.

If 3D printing is given the value of a dream because it is a magical paintbrush, 4D printing is the Monkey King's *Jingu Bang*,[‡] which must be an artifact that everyone is familiar with and longing for, though, it only exists in the novel. But today, the *Jingu Bang* that once existed in the novel has walked into our real-life with the arrival of 4D printing. The only difference is that the new technology is 3D printed and has the function of *Jingu Bang*.

SECTION 3 | 3D PRINTING VS. 4D PRINTING

3D encompasses three dimensions: length, width, and height, which are the three-dimensional world as understood by human senses. However, compared to 3D printing, 4D printing adds an additional dimension, which is time (T). Therefore, we can express the relationship between 3D and 4D printing with the formula: 4D = 3D + Time.

Broadly speaking, based on 3D printing and capable of responding to external stimuli to produce changes in performance or functionality, can be termed as 4D printing. The external stimuli mainly include heat, magnetism, light, humidity, pH, etc. The core of 4D printing technology lies in programmable, designable structures that change over time under specific conditions, and its realization depends on both 3D printing and intelligent materials that meet the functional requirements of 4D printing.

From the perspective of materials, the performance of materials used in 3D printing is stable, and the structural models obtained are static, with their

---

‡ Monkey King is the character of a Chinese novel, *Journey to the West*. *Jingu Bang* is a magical cudgel wielded by Monkey King.

shape and performance not changing over time. In contrast, 4D printing involves mathematically encoding intelligent materials through 3D printing to obtain dynamic structures capable of morphing their shape, performance, and function in response to external stimuli; these are unstable and dynamic.

From the perspective of shape and property changes, 3D printing technology strives to stabilize the shape and performance of manufactured products, minimizing deformation and property changes, whereas 4D printing technology takes full advantage of the phenomena of shape and property changes after manufacturing, allowing products to perform differently based on environmental conditions.

From the design method perspective, 3D printing employs a solid static design, where designers only need to design a product's single shape and performance, whereas 4D printing requires dynamic prediction of the product, not only designing the final shape, performance, and function but also programming the material based on its properties to design intermediate shapes and performances.

In the world of 4D printing, the printing object can automatically transform with a specific time set or under specific conditions, or else, it can be self-transformable and it can even self-assemble. What we print could be a cylinder bar. However, it will self-transform into a three-dimensional square when triggered or self-transform into some other shape.

Imagine that for people who travel frequently, luggage of different sizes is required based on different trip conditions. So, we may have multiple pieces of luggage prepared at home or barely use the luggage that is not suitable for the trip. But when we print luggage through 4D printing, these troublesome situations are gone. The luggage can self-expand and self-shrink by keeping up with the pressure generated by the capacity of the belongings in it, and it can self-transform and self-assemble based on the different forms of the things loaded in it. This is the world of *Jingu Bang* technology brought by 4D printing, and it will soon come into our daily lives. Let's say 3D printing will affect and change business forms, and then the changes and impact caused by 4D printing will be even more far-reaching and profound.

SECTION 4 | THE UPGRADE OF DIGITAL MANUFACTURING

Products manufactured through 4D printing technology rely on computers for all pre-formation tasks, such as data scanning, software design, data modeling, and output program design. These fall under the category of digital manufacturing technology. Unlike other digital manufacturing technologies, this is "direct digital manufacturing" (DDM) meaning that products made with 4D printing are final products or parts that people can use. Moreover, this digitization will be comprehensive, from the construction of creative concepts to the realization of creative designs, relying on AI design software. When it comes to the printing stage, it mainly depends on the output of the digital design model, with the 4D printer materializing the digital model into a physical entity.

WHAT IS DIGITAL MANUFACTURING TECHNOLOGY?

Digitalization refers to converting complex and variable information into measurable numbers and data, then using these numbers and data to build digital models, transforming them into a series of binary codes, and processing them uniformly inside a computer.

Digital manufacturing is the digitization of the manufacturing field, resulting from the intersection, integration, development, and application of manufacturing technology, computer network technology, and management science. It also represents the inevitable trend toward continuous digitization for manufacturing enterprises, manufacturing systems, production processes, and production systems. Its contents include:

- CAD—COMPUTER-AIDED DESIGN
CAD has evolved with the development of computer software and hardware technologies. When it was realized that merely using computers for drafting could not be considered computer-aided design, and that real design should

encompass the entire product, including conception, functional design, structural analysis, and manufacturing, traditional two-dimensional engineering drawing was recognized as only a small part of product design. Thus, CAD evolved from its original meaning of computer-aided drawing to computer-aided design, not just assisting in drawing but helping in creating, modifying, analyzing, and optimizing design technologies.

- CAE—COMPUTER-AIDED ENGINEERING ANALYSIS

CAE typically refers to finite element analysis and the kinematic and dynamic analysis of mechanisms. Finite element analysis can perform mechanical analysis (linear, nonlinear, static, and dynamic), field analysis (thermal, electrical, magnetic fields, etc.), frequency response, and structural optimization. Mechanism analysis can calculate the displacement, velocity, acceleration, and force of components within a mechanism, simulate the motion of the mechanism, and optimize mechanism parameters.

- CAM—COMPUTER-AIDED MANUFACTURING

CAM is an acronym for computer-aided manufacturing, capable of automatically generating the numerical control code for part machining based on CAD models, dynamically simulating the machining process, and completing interference and collision checks during machining. CAM systems, combined with digital equipment, can realize paperless production and lay the foundation for the implementation of Computer Integrated Manufacturing Systems. The core technology of CAM is numerical control technology. Parts are usually represented by points, lines, and surfaces in a spatial Cartesian coordinate system, and CAM involves controlling the movement of tools on a CNC machine according to these digital quantities to complete part machining.

- CAPP—COMPUTER-AIDED PROCESS PLANNING

CAPP refers to the use of computer software and hardware technologies and supporting environments to perform numerical calculations, logical judgments, and reasoning to devise the machining process for parts. With the help of CAPP systems, it is possible to address issues such as low efficiency, poor consistency, unstable quality, and difficulty in optimization associated with manual process design. CAPP also assists technologists in the design and manufacturing process from raw material to finished product using computer technology.

• PDM—PRODUCT DATA MANAGEMENT

PDM is a technology used to manage all product-related information (including parts information, configurations, documents, CAD files, structures, permissions, etc.) and all product-related processes (including process definition and management). Implementing PDM can improve production efficiency, facilitate the management of the entire product life cycle, enhance the efficient use of documents, data, and standardize workflows. Beyond data management, PDM also provides unified and effective management of related market demands, analysis, design and manufacturing processes, all changes throughout, user instructions, and after-sales service data.

These digital technologies were built before the advent of AI generative technology, representing the core technology of the third Industrial Revolution. However, in the era of the fourth Industrial Revolution, led by AI, these digital tools, whether in design, planning, management, or production, will be integrated by AI, becoming simpler, smarter, and more powerful.

THE DIGITAL MANUFACTURING PROCESS OF 4D PRINTING

The application process of 4D printing technology can be understood as a program coding upgrade based on the foundation of 3D printing technology. 3D printing is based on the principle of forming by layering discrete materials, building models based on three-dimensional CAD product data, using software and data systems to layer and solidify specialized materials, and rapidly manufacturing physical products. It is a digital manufacturing process in the field of digital manufacturing technology known as rapid prototyping technology. This technology automatically, directly, rapidly, and precisely materializes design ideas into prototypes with certain functions or directly manufactures parts. This allows for rapid modifications and functional testing of product designs, effectively shortening the product development cycle.

4D printing technology, built on 3D printing, directly programs the manufacturing materials. In the data model design phase, it fully considers factors such as the material's triggering medium and time for deformation, embedding related digital parameters into the printing material beforehand, allowing it to fold internally on its own. After the model is printed and formed, it can deform, reorganize, and disassemble in both time and space dimensions, embarking on its subsequent life cycle journey.

Compared to 3D printing, 4D printing can be described as a higher level of DDM technology. This technology integrates comprehensive and systematic knowledge from multiple disciplines, including computer software, materials, mechanics, and network information. In the software design phase, many things and issues are abstracted, and they are abstracted from different levels and perspectives; problems or things are decomposed and modularized making problem-solving easier, the finer the decomposition, the more modules there are; simultaneously, by writing corresponding program codes, the material's triggering medium is incorporated into them.

Nervous System is a generative design studio that works at the intersection of science, art, and technology. It has been using 3D printing for the Cell Cycle project since 2009.

By Jessica Rosenkrantz, co-founder of Nervous System, Cell Cycle is a web-based app that allows people to design their own, complex, honeycomb-shaped jewelry and sculptures, which can also be sold and 3D printed online.

Jessica Rosenkrantz said that the digital manufacturing process of 3D printing could produce complex and organism forms of objects that couldn't be made by other methods. Additionally, it could also print unique customized products. Examples include 3D printers that can print a necklace without breaking it, not to mention make whistles with peas and bottles with boats. Some 3D printers can even directly print out the operative parts, such as a multi-part gearbox.

From the introduction of Jessica Rosenkrantz, we can see that the applications of 3D printing will eventually lead to the development of DDM. That is, DDM will use 3D printing to manufacture the final products or some of their parts. What's more, DDM has gained popularity in different industries, such as aerospace, jewelry manufacturing, dental technology, toy production, branded furniture design, and the fabrication of customized fashion items like sunglasses. Following closely behind 3D printing, the potentially overtaking technology of 4D printing embeds related digital parameters directly into the raw materials of the items being manufactured. This allows objects printed with 4D technology to undergo shape changes or perform specific functions under certain conditions based on preset parameters. This adaptability introduces new possibilities for the manufacturing industry, especially in fields requiring complex, transformable structures.

For instance, in the medical field, 4D printing technology can be used to manufacture smart medical instruments and transformable medical devices, including creating supports or implants that adapt based on the changing conditions within a patient's body, improving therapeutic outcomes. In architectural design, 4D printing can be used to create transformable structures, allowing buildings to adapt according to seasonal or climatic conditions. Such buildings, capable of changing shape under different conditions, help enhance energy efficiency and environmental adaptability.

The application of 4D printing is like the birth of human beings. Men's sperm and women's eggs are matched and combined to form a brand-new cell, and the "genes" contained on each chromosome store all the information of race, blood type, pregnancy, growth, and process of life apoptosis. When the living individual is born from the mother, they immediately begin to experience changes in birth, growth, disease, age, and death under the influence of the genetic code. This will further deduce the reproduction of life and the physiological processes of cell division and protein synthesis, among other functions, that are repeating in cycles.

SECTION 5 | REALIZING SPATIAL INTENSIFICATION

The application of 3D printing has created great changes to existing techno-logical innovations and industrial practices. Still, it is far from reaching the acme expected by the industrial technology revolution. At this point, what kind of expectations will 4D printing bring to human society?

The creation of a "concept model" based on 3D printing allows products to be displayed in a physical form in the early stages of manufacturing, which greatly improves the accuracy of product design and model-building. Yet, more physical printing has raised strict requirements on the printing space and skills, and it greatly constrains 3D printing from moving toward "popularization." As a result, 4D printing, which has developed and evolved on top of 3D printing, allows product design and model-building to be simplified to the greatest extent so that physical printing is no longer confined to its formed volume and space.

Before then, several buildings built using 3D printing were unveiled in Suzhou, which included a villa with an area of about 1,000 square meters, a five-story residential building, and a simple exhibition hall. The walls of the buildings were printed by a large 3D printer. It only takes one day to print out a house, and the results are amazing. Today, this 3D printed villa which is priced at RMB§ 80 million, only costs over RMB 1 million.

The printer that makes the dream of printing a villa real is a three-story high "giant monster" with a height of 6.6 meters, a width of 10 meters, and a length of 32 meters, of which the foundation is as large as a basketball court. However, the width of the printed material can only be 1.2 meters. Referring to the computerized design drawings and plans, a huge nozzle controlled by the computer ejected "ink" just like it was icing on a cake, and the "ink" was stacked in a Z shape layer by layer. Soon, a high wall was built. After that, the walls were built up like blocks and were connected by infusing with reinforced concrete in the second "printing."

The 3D printing technology makes this dreamy idea happen, but the giant printer and limited-size prints will undoubtedly constrain the applications of 3D printing. At this junction, 4D printing, which possesses developed technology and extended dimensions upon 3D printing, largely solves the problems of 3D printing in mold, volume, and product irregularities incurred during product fabrication.

In the early stages of printing, which is product design, the designer has implanted the raw materials with designed parameters such as trigger medium, time, etc. By doing so, the printed product triggered by a certain medium

§ RMB stands for renminbi, the currency of China.

can obtain transformation and reconfiguration to realize the shaping and restructuring of the product, which will minimize the difficulty of product printing and maximize the possibility of printing products in different spatial dimensions.

Let's build a villa in Suzhou as an example. What would it be like if 4D printing was applied?

First of all, starting from the design phase, by leveraging AI in architectural design software, we can quickly complete the design of houses without relying on professional architectural firms or architects. Instead, AI design software assists us in accomplishing this task. This approach significantly accelerates the design process, democratizes access to high-quality architectural design, and potentially reduces costs associated with traditional design services. AI-driven design tools can analyze vast amounts of data, including site conditions, environmental factors, and user preferences, to generate optimized, innovative, and functional design solutions that might not be immediately apparent through conventional design methods. Second, the printer is no longer the "giant monster" with a big size and aggressive height but a more compact and exquisite printing technology. During the printing process, implanting the higher-tech design parameters allows the printed products to have a transformable ability without the constraint of the 1.2-meter width. The printer can also print out more compact products, which may extend infinitely when triggered by the medium. Just like Monkey King's *Jingu Bang*, with the medium of a spell, "big" or "small," it can become as small as a silver needle or as large as the sky. The 4D printed products can change not only in size but also in shape.

Within a limited space, infinite possibilities are printed out with the use of the design and implantation of different parameters. All these products that have been designed and implanted with various changing parameters in their raw materials will bloom and display in the required time and space dimensions, completing their perfect journey of 4D printing. All the enchanting tech and incredibleness of 4D printing is simple but not easy.

SECTION 6 | MEETING DIVERSE NEEDS

If we want to actualize 3D printing, in addition to printers and materials, we must have 3D models. However, as far as the current situation is concerned, there are still some professional problems in the design of 3D models, and it is difficult for general consumers to produce decent 3D models.

A company called Volumental has in-depth ideas in this regard, mainly aimed at allowing general consumers to complete their desired design of 3D models by use of software. The app provided by the company lets people make 3D models simply and swiftly. A consumer just needs to take a photo of the object that he or she wants to print in 3D with a depth camera, and then transmit the data to the app, and eventually, obtain the 3D model on the app.

The commercial application of the 3D scanning technology provided by Volumental is very widely implemented, such as product customization. Through this process, consumers can pass the 3D data of their feet to the merchant, and with the given data, the merchant will be able to customize their shoes accurately. Similarly, consumers can also scan the 3D data of their entire body to have custom-fitting clothes, or even with their stereo pictures of the same proportions.

Volumental and Intel jointly announced a plan to work together to reform the shoe-selling industry online. The concept is to sell shoes based on the 3D data of consumers' feet, and this is only the first step of their collaboration. In the future, they will further collaborate to create more new business opportunities. Meanwhile, more and more companies have begun to apply the 3D data of consumers' bodies to their commercial fields.

Along with the bigger extension space and more changes generated by 4D printing, there will be more space for the software that auto-produces design depending on the scanning technology.

For instance, users can take two different photos based on their ideas, one of the initial model and another of the model after self-assembly. The design software will start working with scanning technology, automatically analyzing and identifying the differences between the initial and self-assembled models, including changes in shape, size, color, etc. Subsequently, the system software will use this data to generate a design automatically and synthesize the self-assembly process program. This means that we do not need professional design skills; simple photography allows us to participate in the creative process.

Moreover, with special raw materials, 4D printing can achieve folding scans that 3D printing cannot, integrating various demands into the final product through 4D printing technology. Specifically, through special raw materials, 4D printing technology grants manufactured goods the ability to change in the fourth dimension. Compared to traditional 3D printing, this folding scan printing capability offers users greater freedom, making the final product not limited to the initial static design but capable of dynamic changes in real-time according to needs.

4D printing technology can also integrate a variety of different needs into the same final product, achieving a higher degree of customization and comprehensiveness. This brings new possibilities to manufacturing and creative design, making products not rigid but dynamically changeable according to actual needs.

In regard to clothing, in the near future, not only will we be able to customize clothes, but also simultaneously integrate the dressing codes of different social occasions into one piece of clothing through 4D printing. To meet the different needs of social occasions, a consumer can trigger the predetermined instructions of the worn clothes with the respective medium device at any time to let it complete self-transformation and self-assembly in a short while. Such change won't be constrained by time, and for women with a desire for a good appearance, the applications of 4D printing on things like jewelry, shoes, etc., will make their lives more colorful.

SECTION 7 | 4D IN EXPERTS' EYES

If 3D printing is likened to "gathering sand to form a tower," connecting discrete materials with the aid of computer-aided design into an integrated whole with specific shapes and functions, then 4D printing is about using materials with specific properties for additive manufacturing and endowing parts with intricate structures the ability to change shape over time, making them "flexible and extensible." Since its conceptualization in 2013, 4D printing has generated significant market interest. 4D printing adds the dimension of time and designability to traditional 3D printing technology—it can be said that the advent of 4D printing has enriched the application potential and vitality of 3D printing, demonstrating a powerful force for future transformation.

Looking further, 4D printing technology is a combination of 3D printing technology and novel material technology, essentially using 3D printing technology to create substances that change over time according to design.

A report by the Atlantic Council suggested that 4D printing technology not only holds significant economic, environmental, political, and strategic importance like 3D printing but can also "map" digital information from the virtual world (programming information) onto physical objects in the material world. This means that 4D printing possesses not only the advantages of 3D printing but also the benefits that 3D printing does not offer.

Looking at the shared advantages, both 3D and 4D printing can enable manufacturing enterprises to more flexibly respond to market demand changes, achieving personalized and customized production. This will drive a shift in production methods from mass production to small-scale, personalized production, improving production efficiency, reducing waste, and having a significant positive impact on the economy. Additionally, due to the flexibility of printing and the characteristics of personalized production, there is no longer a need for mass production and inventory, reducing the issues of overcapacity and waste products. This helps reduce the environmental burden, promoting the development of sustainable manufacturing, and aligning with modern society's urgent demand for environmental protection.

Looking at the unique advantages of 4D printing, first, 4D printing can achieve a higher level of automation. Traditional manufacturing methods may require manual intervention and adjustments, while the advantage of 4D printing lies in the guidance of programming information, enabling objects to be assembled, repaired, or transformed automatically under specific conditions. This highly automated manufacturing process not only increases efficiency but also reduces the demand for human labor, pushing the manufacturing industry toward a more intelligent and flexible direction.

Second, the mapping of digital information makes the design and innovation of objects more flexible. Designers can directly control the shape, function, and response mechanism of an object through programming information, without the need for tedious manual operations. This provides greater freedom for creative design, allowing products to not only display unique forms in a static state but also dynamically change over time, achieving more diverse and intelligent designs.

From a sustainability perspective, the advantages of 4D printing are also evident. The mapping of digital information makes the manufacturing process more precise and controllable, helping to reduce material waste, as the fabrication of objects can be more finely tailored to demand without the need for large-scale excess capacity.

Furthermore, the mapping of digital information enhances the customization of production. Programming information can be adjusted and modified, allowing the same design model to be customized according to the needs and environmental conditions of different users. This provides greater flexibility for the manufacturing industry, enabling products to better adapt to diversified and personalized market demands.

It can be said that by directly mapping digital information from the virtual world onto specific objects in the material world, 4D printing not only improves the level of manufacturing automation but also increases the flexibility of design and innovation. This advantage helps push the manufacturing industry toward digital and intelligent directions, creating more favorable conditions for innovation and sustainable development.

Currently, 4D printing technology has been preliminarily proven in laboratory environments, and research in various fields is accelerating breakthroughs, including in the food sector, biomedicine, consumer products, and even military applications. Suppose that 4D printed products can be used in the underground piping system, which has been plagued by the factors of fixed capacity and high maintenance costs for a long time. Using programmable materials, each pipeline can adapt to the changing environment, while its capacity and flow rate can be adjusted by self-expanding or self-extracting, even self-repair when damaged or self-decomposition when scrapped. Besides, the article especially pointed out that 4D printing could be used in national defense. The US Army has begun to utilize 3D printing to make new equipment for the frontline in Afghanistan, though, 4D printing can allow the military to manufacture automobiles of which the coating could change its structure to adapt to any terrain. It could also help to make a soldier's uniform that could protect against poisonous gas.

Of course, like many other emerging technologies, 4D printing also has worrying risks and negative effects, especially the potential risks that may arise in its application in the biological field. For example, researchers at Harvard University are already using 4D printing to create nanorobots with DNA strands to fight cancer. Because the tools used in this application can be easily obtained in today's world, some people may use this technology to create new biological weapons. Besides, many risks associated with the wide implementation of 3D printing applications will also be accompanied by the development of 4D printing, such as being used for criminal means (pistols and handcuff keys have been previously produced through 3D printing). In the end, even though 4D printing can provide industrial manufacturers with more ways to customize products so that the supply chain can be further shortened, it will also endanger and complicate the technical work, product accountability, and intellectual property questions.

Indeed, like all new technologies, including AI, 4D printing also comes with its risks and negative impacts. Like many emerging technologies, 4D printing combines various new techniques, methods, and disciplines. Given the current trends in technological development, the most concerning risks are likely to emerge in the biological field. We can use the principles of 4D printing to create nanorobots that fight cancer using DNA strands. We can also use 4D printing technology to manufacture foldable proteins and set specific mutation triggers at the initial stage of printing. Thus, with 4D printing technology, we can achieve precisely targeted therapy but also precisely targeted mutations. In this regard, the dual-use nature brings about genuine concerns. With the help of AI in research and design systems, many non-professionals can easily access the necessary development tools, and some individuals could use this technology to create new biological weapons, which is a matter of concern.

However, whether we look at the trend of technological development or explore the direction of future business, 4D printing, following in the footsteps of 3D printing, offers more foresight and disruptiveness for human development than 3D printing technology. It represents not just a revolution in production tools but also a change in the future commercial ecosystem of society, medical science development models, and human living conditions triggered by the transformation of production means.

It can be said that what 4D printing will disrupt is not just manufacturing technology but the future of humanity as a whole.

2

THE MATERIAL

As mentioned in chapter one, 4D printing can be simply understood as "3D + T," and its realization relies on two critical foundations: mature additive manufacturing technology—what we know as 3D printing and smart materials that meet the functional requirements of 4D printing.

Based on these two crucial foundations, the advancement of 4D printing is continuously progressing in scientific research, yielding substantial results.

SECTION 1 | THE MACHINES OF 4D PRINTING

3D printing is already familiar to us; it connects discrete raw materials into a three-dimensional whole through point-by-point, line-by-line scanning, and layer-by-layer stacking manufacturing logic. Combined with computer-aided structural design, it can directly produce various complex geometric structures, maximizing material savings and reducing subsequent processing.

From the perspective of forming principles, the currently common 3D printing technologies mainly include Stereolithography (SLA) for polymer materials, Fused Deposition Modeling (FDM), Direct Ink Writing (DIW), and Selective Laser Melting/Sintering (SLM/SLS) commonly used for metal materials, among others.

After decades of development, 3D printing can now maturely realize the formation and manufacturing of complex and fine structures of metals, polymers, and ceramics. It has been applied in numerous fields, such as aerospace, biomedicine, mechanical electronics, industrial design, and cultural education.

However, as the demand in application fields continues to expand, the functional designability of static structures or components produced by traditional 3D printing has shown certain limitations. For example, intelligent devices with self-folding structures from micro to macro scales, considered to have good application prospects in biomedicine, aerospace, and other fields, are difficult to achieve through traditional 3D printing of materials.

Naturally, scientists have thought of using smart materials with shape memory effects, self-healing materials, or materials with special topological structure designs known as "metamaterials." This led to the birth of 4D printing.

This shows us that 4D printing is built on the foundation of 3D printing. 4D printing also involves creating single-use structures by appropriately combining various materials. The different properties of materials, such as expansion rates and coefficients of thermal expansion, will cause the printed structures to undergo the desired shape transformation under external stimuli. Therefore, 3D printers are also essential for manufacturing 4D prints. 3D printing devices usable for 4D printing include FDM printers based on material jetting technology, Stereolithography printers based on photopolymerization technology, and Selective Laser Sintering/Melting printers based on powder fusion technology.

SECTION 2 | SMART MATERIALS

Similar to 3D printing, 4D printing is an advanced manufacturing technology that highly integrates material science, mechanical science, computer science, and many other disciplines. Unlike 3D printing, 4D printing is a smart manufacturing technology with dynamic evolution capabilities, combining smart materials with intelligent structure design on the basis of 3D printing, aimed at further development in the shape, performance, and functionality of manufactured products. Among these, smart materials are the key to the "difference" between 3D and 4D printing. It can be said that smart materials are the most crucial part of the realization of 4D printing technology.

In the late 1980s, inspired by certain capabilities possessed by organisms in nature, scientists in the US and Japan first introduced the concept of intelligence into the field of materials and structures, proposing the new concept of smart material structures. Smart material structures, also known as sensitive structures, broadly refer to integrating sensing elements, actuating elements, and related signal processing and control circuits within material structures. Through mechanical, thermal, optical, chemical, electrical, and magnetic stimulation and control, these structures not only have the capacity to bear loads but also possess various functions such as recognition, analysis, processing, and control. They can perform self-diagnosis, self-adaptation, self-learning, and self-repair.

Smart material structures are a cross-disciplinary frontier subject, involving a wide range of professional fields such as mechanics, material science, physics, biology, electronics, computer science and technology, etc. Smart materials used for 4D printing can be divided into shape-changing materials (SCMs) and shape-memory materials (SMMs).

SCMs act as a switch that immediately changes shape upon external stimulation and returns to its original state once the stimulus is removed, such as materials that expand or shrink linearly in volume.

SMMs are characterized by their structure deforming under external stimuli, putting the material in a temporary deformed state. In this state, the material "remembers" its shape, and upon receiving external stimuli again, it can revert to its original form. Materials with shape memory characteristics include Shape Memory Hydrogels (SMH), Shape Memory Ceramics (SMC), Shape Memory Alloys (SMA), Shape Memory Polymer Composites (SMPC), and Shape Memory Polymers (SMP), with Shape Memory Polymers being the most extensively researched category.

SMH

SMH is a type of polymer material with adaptive functions, capable of temporarily deforming in response to various external stimuli and permanently remembering its original shape. As a novel smart material, SMH's memory function is achieved through two special cross-linking structures within its three-dimensional network: permanent cross-links and reversible dynamic cross-links. This material can be categorized into pH-responsive SMH, electrically-responsive SMH, and temperature-responsive SMH, based on different response mechanisms.

pH-responsive is a unique response mechanism of gel materials, working based on the material's internal hydrogen ion concentration responding to different pH values. Researchers have previously created SMH that responds to specific pH conditions in various shapes using photocatalytic methods. Electrically-responsive SMH changes its macroscopic shape by creating an ionic concentration difference between the hydrogel and its solution due to changes in the electric field. In 2018, researchers at Rutgers University in New Jersey demonstrated the feasibility of electrically-responsive SMH by exploring how changes in ionic concentrations in hydrogel solutions cause shape changes. The high water content three-dimensional network within SMH contributes to its high hydrophilicity. Moreover, the advantages of SMH as a 4D material include strong self-healing, low cost, and high biocompatibility.

SMC

SMC is a rigid and hard material capable of withstanding high working temperatures in harsh environments with almost no strain. In 4D printing, liquid ceramic suspensions are often introduced through magnetic or thermal treatments.

Specifically, one application of liquid ceramic suspension is achieving anisotropic shrinkage through magnetism. That is, magnetic particles can be introduced into the ceramics, and then, using a magnetic field, the ceramics can be made to shrink in a specific direction. This method can be used with shape-programmable polydimethylsiloxane-based nanocomposites to achieve shape changes in ceramics under specific conditions. Another application of liquid ceramic suspension is shape programming during thermal treatment, i.e., introducing specific nanocomposites into the suspension to cause shape changes in ceramics when heated. This can be achieved by designing and controlling the materials in the liquid ceramic suspension.

Among the currently available liquid ceramic suspensions, ceramics based on zirconia (ZrO_2) have attracted widespread attention. This is because these ceramics exhibit reversible martensitic phase transitions similar to SMA. Meanwhile, other ceramics, such as multiferroic perovskites, also exhibit shape memory behavior through reversible martensitic phase transitions, driven by external electric and thermal fields, primarily based on ferroelectric or piezoelectric properties.

In terms of applications, due to ceramics' special properties, liquid ceramic suspensions can be used to achieve high actuation stress and strain, as well as a wider range of transformation temperatures.

In ceramic shaping, there is often a need for cost-effective processes to prepare complex geometric shapes. Inspired by the organization of cellulose microfibrils in plants, some research attempts to program the microstructure of ceramics to undergo local anisotropic shrinkage during thermal treatment. This method achieves self-shaping, where the design of the microstructure can control local shrinkage, for example, by arranging alumina-reinforced thin films. Compared to the current mechanical machining and injection molding processes used for ceramic shaping, this bio-inspired method reduces waste production and eliminates the need for expensive mechanical equipment. Moreover, this method allows for the manufacture of ceramic parts with complex geometries without weak interfaces or joints, offering new possibilities for the cost-effectiveness and design flexibility of ceramic manufacturing.

SMA

SMA is a type of smart metal material with a memory effect. It stands out as one of the materials for 4D printing due to its unique shape memory characteristics, variable elastic modulus, super elasticity, high damping, and excellent biocompatibility—qualities that are difficult for conventional metal materials to possess simultaneously.

SMA mainly achieves the reversible transformation between austenite and martensite phases through thermal or magnetic stimuli. These alloys have a unique memory effect, with thermal and magnetic responses being the two main driving mechanisms.

In thermally responsive SMAs, the memory effect is caused by the reciprocal change from the austenite phase at high temperatures to the martensite phase at low temperatures. This process involves thermoelastic martensitic transformation and its reverse transformation, leading to macroscopic shape changes in the material. For example, the Ti-Ni series of memory alloys, which exhibit thermoelastic martensitic transformation, is one of the most widely used SMA. Simply put, this is like the material having one shape at high temperatures and then automatically returning to another shape upon cooling through a special phase transition process.

Previously, researchers have utilized the deformation characteristics of $Ti5_0Ni5_0$ (atomic percent) at different temperatures and its excellent biocompatibility to create spiral vascular stents using 4D printing technology. In a low-temperature environment, the stent contracts, allowing it to be implanted into blocked blood vessels through minimally invasive surgery. Influenced by body temperature within the blood vessel, the stent expands to support the vessel, allowing blood to flow normally through the blocked vessel.

The memory effect of magnetically responsive SMAs is due to micro-level twin boundary sliding and the reorientation of martensite phases in the direction of the magnetic field, causing macroscopic shape changes in the material. This process is like the material being controlled by a magnetic field, achieving overall shape changes through minute structural modifications.

First, micro-level twin boundary sliding refers to the movement and adjustment of tiny grains within the material when stimulated by a magnetic field. This sliding leads to minor overall deformations of the material, laying the foundation for subsequent macroscopic shape changes. Second, the reorientation of martensite phases in the direction of the magnetic field occurs when the martensite phases rearrange themselves to align with the direction of an external magnetic field. This process results in overall macroscopic deformation of the material, achieving shape changes controlled by the magnetic field.

It's worth mentioning that, unlike other smart shape materials, SMAs possess excellent mechanical properties typical of metal materials.

SMPC

Distinct from other categories, SMPC resides in the overlapping area of SMM, where at least one SMM forms part of the composite materials as a monomeric unit, and each monomer plays a crucial role in the final design. SMA and SMP are two common types of shape memory materials, each displaying different shape recovery mechanisms and widely studied across multiple fields for their shape memory effects. However, these SMMs have their respective shortcomings, such as the high cost and low strain recovery of SMA.

To overcome these issues, scientists have combined SMA and SMP to create SMPCs. For instance, some studies have used 3D printing technology with nylon 12 as the filament material to manufacture reversible SMPC-driven 4D printed actuators. They optimized the volume fraction ratio of SMA to SMP to find the best operational cycle ratio for applications in supports and valve controllers in additive manufacturing. This combination of SMPCs leverages the advantages of both SMA and SMP, overcoming their individual drawbacks and offering more possibilities for printing designs.

SMP

SMP is a type of polymer material with a memory effect and is currently the most diverse and widely applied shape memory material in 4D printing. Depending on the stimulation mechanism, it can be divided into thermally induced SMP, light-induced SMP, and electrically induced SMP.

Thermally induced SMP's memory characteristic stems from the incomplete compatibility of molecules inside the material. After being printed, a thermally induced SMP will maintain one state at room temperature and transition to another pre-programmed shape upon uniform heating. By printing 3D objects from materials with different dynamic mechanical properties according to custom designs, objects that can change shape in a controlled, predetermined sequence can be created. Through 4D printing technology, the material's response to temperature can be pre-programmed, enabling complex structures to fold automatically. When heated, each type of SMP changes shape at different rates, depending on its internal clock. By precisely designing the sequence of these material shape changes, the self-assembly of three-dimensional objects can be achieved.

As early as 2004, the Keystone Research Group made significant progress by wrapping SMP around a wick. This material can change from rigid to elastic and flexible under thermal stimulation, then return to its rigid state. By adding high-strain fibers to the material, its toughness can be enhanced.

Currently, research on thermally induced SMP primarily focuses on epoxy resins and polyurethanes. In the 20th century, Mitsubishi first developed polyurethane SMP, and subsequent research adjusted the composition and ratio of components to successfully create shape memory polyurethanes with different response temperatures. Additionally, CTD Corporation in the US has developed a series of thermosetting epoxy resins with memory effects, capable of good shape recovery even in extremely low-temperature environments.

Simply put, the memory characteristic of thermally induced SMP fully utilizes the incomplete compatibility of internal molecules. Stimulated by external temperature changes, these materials can reversibly change shape between rigidity, elasticity, and flexibility, offering a wide range of possibilities for practical applications. By introducing different components and adjusting ratios, more diverse response temperatures and shape memory effects can be achieved.

Light-induced SMP utilizes photosensitive ions to absorb light energy, converting light energy into chemical energy to raise the material's temperature.

When the temperature reaches the response value, the shape memory effect is triggered. Studies have shown that using multicolored SMP composites can achieve structural changes under different lighting times and colors, thus enabling remote light-driven activation. Compared to thermally induced SMP, light-induced SMP is more flexible and can be selectively driven in specific areas based on the position and direction of the light source and photosensitivity differences.

Electrically induced shape memory polymers are made by filling SMP materials with conductive particles to form a conductive network, granting SMP conductivity. The heat generated by an electric current triggers the material to deform based on temperature changes, thus exhibiting shape memory characteristics. Researchers have previously explored the volume resistivity and electrically-driven memory characteristics of CNT/PLA composite materials made from carbon nanotubes (CNT) and polylactic acid (PLA) under the influence of direct current, investigating how temperature changes affect the material. Experiments have shown that printing speed, layer thickness, and raster angle significantly influence electrically-driven memory behavior.

The self-folding process of smart shape-memory materials respond slightly differently to temperature. Materials with subtle differences in folding rates are used to ensure that components do not interfere with each other during folding. The components can react to stimuli such as temperature, humidity, or light in a precisely timed manner, thus forming three-dimensional structures, deployable medical devices, robotic toys, and various other configurations.

In layman's terms, the material used in 4D printing is based on a type of "shape memory" material, or rather, a material with intelligent logic. However, materials that currently meet the requirements of 4D printing technology are still quite scarce. If the development of materials for 3D printing is still on its way, then materials for 4D printing are just at the beginning stage. Yet, the emergence of this technology will inevitably trigger a revolution in new materials and promote the development of new materials. Materials based on 4D printing will be a significant opportunity in the future, with commercial value comparable to that of the "gold mine" of big data.

SECTION 3 | MAKING THE PRODUCTS "WAYWARD"

We already know that 4D printing can achieve the response of components to changes in the external environment by using smart materials with different properties. Besides smart materials, another important part is the change in the external environment, that is, the stimulus to the material. Depending on the response of the printed parts to external stimuli, 4D printing can be driven by water, heat, magnetism, electricity, light, and other methods.

WATER-DRIVEN 4D PRINTING

Materials that respond to water typically expand upon contact with water. Such materials are not uncommon, similar to the water-absorbing layer found in soybeans, "water babies," and diapers. By cleverly utilizing these common materials, researchers have successfully achieved the deformation of 3D printed structures underwater, which is water-responsive 4D printing.

The key to water-driven 4D printing lies in the expansion properties of the material, but the expansion rate of a single material is the same. If only materials with a single expansion rate are used for printing, the structure in water will only "enlarge" and "shrink" simply, without achieving complex deformation. To solve this problem, researchers have used a multi-material printing method, similar to a sewing machine, weaving two or more materials together in a specific way. Through this method, the different arrangements and bending directions of materials enable the entire structure to achieve complex deformation.

The same material can also achieve 4D printing. In 2018, the academic journal *Nature Materials* reported such a breakthrough. In this study, scientists adopted an innovative approach by adding nanofibers to the solution, creating a special paste suitable for DIW printing. This special paste exhibits shear-thinning properties when passing through the nozzle, becoming more fluid. The nanofibers realign along the direction of extrusion during this process. In this state, the cured paste exhibits anisotropy, meaning the same material produces different expansion rates when the expansion rate perpendicular to the extrusion direction is higher than parallel. Using a method similar to DIW, scientists successfully printed a double-layer structure, where the same material exhibited different expansion rates in different directions, achieving complex deformation and forming a unique changing structure in water.

Water-driven 4D printed components typically consist of a driving component that changes volume when exposed to water and a hydrophilic material as the matrix component. When the driving component combines with water molecules, it changes volume and deforms. For example, hydrophilic polymers form hydrogels when encountering water, causing a rapid increase in volume, and cellulose expands when combined with water molecules. The deformation of the matrix's driving components ultimately causes the overall structure in the water environment to deform. The main considerations for achieving water-driven 4D printing are preparing printing materials with anisotropic swelling and designing printing materials with different swelling characteristics in different directions in a water environment.

Water-driven 4D printing materials are relatively easy to manufacture without complex printing equipment. It can achieve significant deformation and is expected to be applied in fields like the human body and underwater robotics. However, since components made of water-responsive smart materials are highly dependent on the water environment, achieving precise remote control remains a challenging issue.

THERMALLY-DRIVEN 4D PRINTING

The materials and processes for 4D printing with thermally responsive materials are relatively mature. These materials are based on phase changes or glass transition within the material, corresponding to the phase change temperature (Tm) or the glass transition temperature (Tg).

In 4D printing, materials are shaped using 3D printing above the phase transition temperature or glass transition temperature. Then, external forces are applied below the phase transition temperature to deform and reconfigure, resulting in a temporary shape. This temporary shape can be maintained below the phase transition temperature. When the temperature returns above the phase transition temperature, the shape reverts to its original form during printing. Currently, a wide range of thermally responsive materials, such as

SMMs, hydrogels, and liquid crystal elastomers, are available for 3D printing. In particular, thermoplastic SMPs or SMAs that respond to thermal stimuli are increasingly used as materials for thermally-driven 4D printing.

Among these, thermally-driven SMPs are easier to prepare than SMA. Hence, they are widely used in research on thermally-driven 4D printing technology. The shape memory function of thermally-driven SMP originates from the glass transition or melting transition of their molecular chain components under temperature stimuli.

The classic manufacturing process for thermally-driven 4D printing components is as follows: first, components with an initial shape are manufactured using additive technology; then, when the component's temperature is above the polymer's glass Tg, the component is adjusted from its initial shape to a temporary shape, which is then cooled below the glass transition temperature to stabilize the temporary shape; when heated again above the glass transition temperature, the component can revert to its original shape, achieving the shape memory function.

In one interesting study, researchers utilized the thermal strain differences between two materials to achieve bending deformation perpendicular to the printing fibers through different thermal strains generated upon heating. This mechanism led to the creation of a unique gripping device composed of a "palm" and "fingers." The "palm" is a pre-fabricated structural component with three flexible "fingers" attached. When the entire structure is heated, the three fingers bend due to the thermal strain differences, simulating the action of grasping an object.

Moreover, when the temperature changes, thermally-driven 4D printing components usually deform overall, but in practical applications, localized deformation of parts is often required. For this purpose, researchers have studied components that can achieve localized deformation. For example, in the design of petals, SMP with different glass transition temperatures was used to achieve the flower's layered blooming and simple control.

Thermally-driven 4D printing components can not only stably maintain a temporary shape but can also adjust the temporary shape by controlling temperature, thereby endowing the components with different mechanical properties. Some researchers, taking advantage of the temporary shape-adjustable characteristics of SMP, have created smart metastable metamaterials with adjustable shapes, significantly variable elastic moduli, and reusability by designing the microstructure of metamaterials. This material can be applied in areas such as soft robotics and morphing wings.

Magnetic-Driven 4D Printing

Magnetic-driven 4D printing technology activates and controls 4D printed components through a magnetic field. There are two main implementation methods: direct response and indirect response.

The direct response method involves embedding magnetic particles within a matrix to fix a temporary shape and then placing it in a magnetic field. The magnetic field changes the magnetic domains within the magnetic particles. When the same magnetic field is applied again, the magnetic field within the matrix's magnetic particles responds to the applied magnetic field, thus realizing shape memory. The indirect response method is based on the magnetothermal effect of magnetic particles in a magnetic field, using heat to drive components. This method is a variant of the thermal-driven method, i.e., achieving controllable shape changes through the material's thermosensitivity.

In the field of magnetic-driven 4D printing, scientists have conducted in-depth exploration. Researchers have used binder jetting 3D printing technology to successfully achieve additive manufacturing of Ni-Mn-Ga magnetic shape memory alloy mesh parts, providing strong support for the application of magnetic-driven technology in the metal field. On the other hand, researchers have also studied composite printing ink materials containing soft magnetic iron particles. Butterfly samples made from PDMS/Fe ink rapidly flapped in an externally applied magnetic field, demonstrating the broad prospects of magnetic-driven technology in biomimetics and biomedical fields. Additionally, researchers used UV-curable gel materials dispersed with magnetic powder and applied a magnetic field to create magnetic anisotropy, manufacturing worm-type soft drives and artificial cilia, showcasing the potential applications of magnetic-driven technology in robotics and nanotechnology.

Compared to thermally-driven and water-driven methods, magnetic-driven 4D printing technology relies less on the external environment and achieves remote control, inducing contactless thermal deformation. This characteristic gives magnetic-driven technology unique advantages in special environments or scenarios requiring specific conditions. Moreover, since the magnetic field can be quickly changed and switched, magnetic-driven 4D printing components usually have a higher response speed, offering possibilities for real-time deformation and precise control.

Electrically-Driven 4D Printing

Electrically-driven 4D printing primarily utilizes the resistive heating effect of electric currents. Simply put, special materials capable of being heated by electric currents, such as heating wires or conductive fillers, are embedded within the material. When these materials are energized by an electric current, they activate the shape memory effect, thereby controlling the shape change of the material. Compared to other driving methods, electrical driving offers higher heating efficiency and quicker response speeds, without the need to alter the external environmental temperature.

In electrically-driven 4D printing, researchers have utilized the FDM principle and carbon fiber-reinforced polylactic acid shape memory composite materials (CFRSMPC) to manufacture electrically-driven 4D printing components successfully. The carbon fibers in this composite material not only act as reinforcing agents but also exhibit significant electrothermal effects.

Using the FDM principle, scientists created a unique composite material by mixing carbon fibers with polylactic acid. The carbon fibers serve as reinforcements, providing strong support for the overall structure. At the same time, these carbon fibers possess an electrothermal effect, meaning they can be heated by electric currents, thereby activating the shape memory effect. Experimental results showed that within just 75 seconds, these components demonstrated an outstanding electrically induced shape memory effect, with a shape recovery rate exceeding 95%.

This result means that scientists can precisely control the shape of the composite material through electric currents without the need to change the external environmental temperature. This opens up new possibilities for 4D printing applications, especially in fields requiring rapid response and precise shape changes, such as medical devices and smart materials.

LIGHT-DRIVEN 4D PRINTING

Light-driven 4D printing technology utilizes light as the activation source to change the structure or appearance of 4D printing components under illumination. A previous study showed that polymers containing cinnamyl groups could deform and fix into predetermined shapes, such as stretched films, tubes, arches, or spirals, under ultraviolet (UV) light exposure. Interestingly, these deformations could remain stable for a long time even when heated to 50°C. Moreover, these polymers could revert to their original shapes at room temperature after exposure to different UV light wavelengths.

Another study involved the use of Near-Infrared (NIR) light-activated SMP, synthesizing V-fa/eco polymers from biomaterials. This copolymer could be used directly as printing material without the need for other polymers as matrix materials. In this research, when the mass fraction of ECO in the printing material exceeded 50%, the printed samples could be remotely driven under 808 nm NIR light within 30 seconds and exhibited a high recovery rate.

One significant advantage of light-driven 4D printing technology is the ability to achieve precise remote control. However, it is important to note that this driving method may face the risk of failure when the components are obstructed or the printing material's transparency is poor, thereby limiting its application scope.

4D PRINTING AIDS ROOM-TEMPERATURE SUPERCONDUCTIVITY

On July 22, 2023, South Korean scientists published a paper on room-temperature superconductivity, claiming to have discovered the world's first room-temperature superconductor, LK-99. This material, named "modified lead phosphate crystal structure (LK-99, a lead phosphate doped with copper)," has garnered worldwide attention. Clearly, this is a highly controversial and epoch-making study, also currently difficult to be confirmed and verified in a short period. Following the publication of this paper, I have publicly commented, mainly expressing two views. First, there are significant challenges in making room-temperature superconductivity a reality, as this research carries considerable uncertainty, and reproducing the results is also challenging. Second, the material approach of the South Korean scientists is very valuable, which involves using the crystal lattice changes on the material's surface to eliminate resistance. This has a significant connection with 4D printing, or to say, 4D printing could play a crucial role in the realization of room-temperature superconducting materials. I will continue to discuss this issue below.

The characteristic of superconductivity requires extremely low temperatures and high pressures to manifest in current superconducting materials, with the market divided between low-temperature superconductors and high-temperature superconductors, the latter being considered "high-temperature" relative to the former, despite still requiring conditions far below freezing. The discovery of high-temperature superconducting materials in 1986 marked a significant advancement, enabling the use of cheaper liquid nitrogen and expanding superconducting applications. Over three decades, China has progressed to the second generation of high-temperature superconducting tapes, achieving mass production after a prolonged development period.

Returning to the South Korean study, the team mixed powders containing lead, oxygen, sulfur, and phosphorus, which underwent a chemical reaction under high heat to form a copper-doped lead phosphate crystal, LK-99. Upon measuring the resistance of LK-99 samples at various temperatures, they observed a drastic reduction in resistivity around 30°C, indicating superconductivity within this temperature range.

The superconductivity in LK-99 is believed to be caused by minor volume contraction (0.48%) leading to structural deformation. As the material cools from high temperatures, its crystal lattice changes, shortening the distance between lead atoms and enhancing electron coupling. This electron coupling allows electrons to flow freely without energy loss, overcoming repulsion and pairing up, thereby achieving superconductivity.

In essence, room-temperature superconductivity could be achieved if the internal crystal lattice of a material can undergo displacement changes, allowing electrons to flow without colliding with the lattice, thus reducing resistance.

4D printing technology could enable material lattice changes at room temperature under specific stimuli, achieving room-temperature superconductivity. Furthermore, 4D printing can tailor materials to undergo specific deformations in certain environments and shapes, thereby realizing superconductivity. This groundbreaking approach offers a new pathway to achieving room-temperature superconductivity, leveraging 4D printing to manipulate material structures for desired electronic properties.

SECTION 4 | NEW DESIGN SOFTWARE

With 3D printers, any imagination can be printed into reality, and with 4D printing, our ideas are endowed with even more variation and imagination. The development of 3D printing technology will inevitably benefit the exploration and application of 4D printing technology, including the advancement of design software. Because 4D printing considers not only the three-dimensional shape of an object but also introduces a fourth dimension, time. This means the printed object can change shape under specific conditions, making it more adaptive and intelligent.

In this context, the innovation of design software becomes crucial to meet the demand for control over the time dimension. For example, if we want to print a wardrobe, it is in a stacked layer state (Figure 2-1) in its original condition. When the user places it in the designated location and provides the triggering medium, the stacked layers automatically drive, change, and assemble into the designed wardrobe (Figure 2-2). In this process, besides the material itself, the self-assembling model becomes the core key. The model connects two phases: the initial phase and the assembly result phase. The driving and connecting roles between are played by the model design process, where the self-assembly process path and outcomes are embedded and then driven by the medium.

The model design clearly requires the aid of computer-assisted software for realization. Therefore, in the development process of 4D printing technology, the further development of CAD software is inevitable, moving beyond the current limitation of static three-dimensional data and driving a more integrated design of models considering the time variable. Perhaps, this process needs to incorporate AI technology.

Figure 2-1 Stacked layer

Figure 2-2 Wardrobe

Specifically, first and foremost, 4D printing design software needs to possess advanced modeling capabilities. Compared to traditional 3D modeling software, 4D design software needs to be able to capture and embed time elements. This includes modeling the shape, structure, and properties of objects at different time points, providing a foundation for the temporal evolution of objects.

Second, the design software needs to incorporate intelligent algorithms and simulation technology. In 4D printing, an object's deformation is often influenced by external stimuli (such as temperature, humidity, light, etc.). This requirement poses a greater challenge for design software because the deformation of an object might occur under the interactive influence of various external conditions. To ensure that objects can change as expected under various conditions, design software needs to implement highly complex simulations. This means the software must accurately simulate and predict material responses and the detailed changes in object deformation under different environmental conditions.

Design software that introduces intelligent algorithms and simulation technology will not just be a simple modeling tool but an intelligent system capable of understanding and adapting to changes in the external environment. Such software will be able to automatically adjust and optimize designs according to specific 4D printing tasks, adapting to different working conditions and requirements.

Furthermore, 4D printing design software needs to provide an intuitive user interface, so designers can easily adjust the temporal evolution of objects. Clearly, if 4D printing is to become widespread and enter everyday life, the professionalism of design needs to be eliminated. That is, any user should be able to input their ideas through a model design platform, and the design software should be able to devise different self-assembly deformation plans accordingly. Therefore, 4D printing design software will need to set parameters such as the start and end times of deformation, adjusting the speed and magnitude of deformation. A user-friendly interface will help encourage more people to participate in the 4D design process, thereby promoting innovation and development.

The development of 4D printing design software also needs to be in tandem with hardware devices. The collaborative work between design software and 3D printers will be a key factor in the future. The software needs to be able to communicate with the printer in real-time to ensure that objects remain synchronized during the printing and deformation process. This requires the design software to have advanced connectivity and communication technologies for efficient interaction with hardware.

It can be said that the design software for 4D printing will become another key technology following printing materials. To seize this wave of opportunity and fully apply and promote this new technology, Autodesk's R&D team has specifically designed a new software, Cyborg, leveraging computer simulation technology principles. It is based on the principles of self-assembly and programmable materials for simulation design, helping users achieve design optimization and handling of material folding relationships.

The magic of 4D printing is distinctly different from 3D printing. The innovation of 3D printing lies in the technology of printers and materials, while the innovation of 4D printing lies in model design or programming, as well as materials, with the latter not being on the same level as the former. The logic of 3D printing is to model in advance and then print the final product through the printer. However, the logic of 4D printing is entirely different. 4D printing embeds product design and time factors into deformable smart materials through a 3D printer, instead of simply presenting the model through printing. Under specific triggering mediums, these smart materials are activated, allowing the printed model or material to self-assemble.

This process does not require human intervention for assembly, nor does it require the implantation of electronic components or electromechanical devices during the model printing process. It is a purely material-based self-assembly and transformation triggered by specific stimuli, ultimately constructing the model originally set by the user.

In other words, the initial model printed by 4D printing might be a flat panel (Figure 2-3), but with the activation by a specific medium, it can automatically self-assemble into a table or a chair (Figure 2-4) within a specific time, without the need for manual labor or any external tools, completely based on the material's inherent design for self-assembly. It can be said that 4D printing not only realizes the assembly of products through the autonomous behavior of smart materials but also endows products with the marvelous function of shape transformation, which is exactly the charm of 4D printing that revolutionizes traditional manufacturing, commerce, and services.

Figure 2-3 Plastic panel

Figure 2-4 Plastic table and chair

This flexible and intelligent manufacturing method also brings obvious advantages. First, it reduces the dependence on human labor, increases automation, and lowers production costs. Second, it eliminates the need for complex external equipment, simplifying the production process and improving production efficiency. Most importantly, the autonomous behavior of smart materials provides a broader space for innovation and personalization. The manufacturing industry can more flexibly respond to market demand changes, producing more unique products that meet individual user needs.

SECTION 5 | 4D PRINTING, SPRINTING FORWARD

As an emerging technology, 4D printing is currently in a rapid development stage. Internationally, research on 4D printing technology mainly revolves around the expansion of 4D printing materials, innovation in molding technology, and the development of design tools, achieving certain results.

First, the variety of 4D printing materials continues to expand, offering more diverse functionalities. New materials are one of the key elements in 4D printing, where materials used must be capable of self-deforming, self-assembling, and adapting under changing environmental conditions. Thus, the expansion of materials has become one of the important research directions in 4D printing technology. After years of development, the types of materials for 4D printing have expanded from polymer materials such as hydrogels to composite materials, organic materials, and ceramic materials.

In 2014, researchers at the MIT developed more types of 4D printing materials, including wood, carbon fibers, textile composites, and rubber. These materials further expanded the application range of 4D printing technology.

In August 2018, a research team from the City University of Hong Kong developed a new type of "ceramic ink" using polymers and ceramic nano-particles. With this, they printed flexible, stretchable ceramic precursors, overcoming the usual deformation limitations of ceramic precursors. Finally, under heat treatment, durable ceramics were obtained, realizing ceramic materials' 4D printing and the manufacturing of complex folding structures in ceramics for the first time. This new technology, characterized by low cost, high mechanical stability, and autonomous deformation, holds promise for application in aerospace propulsion parts, space exploration equipment, electronic devices, and high-temperature microelectromechanical systems.

Moreover, continuous innovation characterizes 4D printing molding technology. In 4D printing technology, using appropriate methods to make the printed smart materials capable of self-assembly, self-healing, and self-deformation as designed is an indispensable link. The constant expansion of new materials brings more challenges in terms of molding, while advanced design technology requires the synchronous and precise printing of multiple materials. Therefore, the innovation of molding technology has become a hot issue in 4D printing research.

In August 2016, MIT utilized micro-stereolithography printing technology to achieve 4D printing of deformable materials at the micrometer scale (1 millimeter = 1,000 micrometers, roughly the diameter of a strand of hair) for the first time. The printed products could recover their original shape within seconds under appropriate temperatures, even after being subjected to extreme pressure or twisting. The researchers used this technology to create a small gripping device that remains open at room temperature and closes when heated, thereby achieving a gripping function. Simply put, the smaller the scale of production, the quicker these materials can recover, with recovery speeds up to just a few seconds. For example, a flower in nature can release all its pollen in milliseconds because its triggering mechanism occurs at the micrometer scale.

Meanwhile, researchers have been searching for the ideal polymer combination to create a shape memory material suitable for their lithography patterns. They selected two polymers, one shaped like curly spaghetti and the other crisscrossing like scaffolding at a construction site. Mixing these two polymers together formed a new material that could withstand significant stretching and twisting without damage. This novel composite material can endure substantial deformation, stretching up to three times its original shape, which is greater than any existing printable material. The new material can "snap back" to its original shape, just as it came out of the printer. Exposed to temperatures between 40°C to 180°C, it can swiftly return to its original form within seconds. If the entire manufacturing process could be scaled down even further, we might be able to reduce the recovery time to just a few milliseconds. This technology has the potential for future applications in aerospace components, solar cells, biomedical devices, and more.

In June 2018, researchers at Virginia Tech developed a multi-material programmable additive manufacturing technique with integrated resin delivery, capable of on-site mixing, delivery, and transformation of resins, as well as self-cleaning. This technique enables micro-scale multi-material additive manufacturing and avoids cross-contamination between different materials, paving the way for 4D printing development at the micro-scale.

From the perspective of design software innovation, 4D printing technology directly incorporates design into materials, simplifying the manufacturing process from design concept to physical object. However, this manufacturing approach also presents new challenges for design work, as designers need to predict the materials' responses under different conditions and base their design work on these predictions. Therefore, 4D printing software emerged. Autodesk developed a design tool software named Cyborg, which can be used to optimize 4D printing designs. This software, through coupled software and hardware tools for simulation, replaces the traditional simulation software's model of first simulating then building or first building then adjusting simulation, enabling simulation of actual deformations in the 4D printing process and allowing users to create a specialized design platform for optimized design.

MIT's Computer Science and Artificial Intelligence Laboratory developed software called Foundry, which helps designers allocate different materials to different parts of a 3D digital model according to design needs, easily achieving multi-material 3D printing and supporting the design work for 4D printing.

Under a series of technological breakthroughs, products with "self-awareness" are being born. It will thoroughly change the current 3D printing and its related industries, such as construction, furniture, piping, clothing, toys, military, etc. Hence, compared with the current hot 3D printing, 4D printing will have broader development prospects and influences.

A good example is IKEA furniture, which focuses on its design and convenient assembly. It has obvious advantages in transportation and assembly compared with traditional furniture, yet we still need to spend labor on assembling it. However, the situation is completely different when it comes to 4D printing. 4D printing is not only more intelligent, but it can print objects that have the ability of "self-creation." What we buy may be just a piece of the multi-layer board but when shipped to the designated location and within the specified time, by being exposed to the trigger medium, which could be water, gas, sound, etc., it will be triggered to self-assemble into a shape (Figure 2-5) without human labor or external force.

Skylar Tibbits believes that 4D printing can, based on 3D printing, give the printed objects the ability to change their shape according to different environmental factors (such as sound, light, heat, water, etc.). Skylar said that the key to 4D printing is the materials that are already known and can change in response to different environmental factors, like memory alloys. By combining these materials with how we want the object to change, we can finally print it out. The ultimate goal of 4D printing is to realize the science fiction scene discussed above. If we buy a set of furniture at IKEA, we don't need to assemble it, and it will self-assemble by itself.

Figure 2-5 Self-assembling furniture

The significance brought about by 4D printing has both an impact on the manufacturing and commercial ecological chains. It will pre-set the result of targeted assembly or molding in the primary model; then the printed object could be triggered by external stimuli within the defined period to perform self-assembly. In the future, it will not only be self-assembled but will also evolve into a form similar to self-fabrication. We may also understand this in a way that 4D printing could implant a kind of memory wisdom into the material, and with the impact of the medium, memory could be triggered to realize the object's assembly.

In this way, maybe one day, we can see a scenario whereby the 4D printed objects could be like robots with no wires or motors to help humans complete projects that are dangerous or beyond human strength. This could include the installation of equipment in outer space or deep-sea or the construction of skyscrapers. The way that the 4D printed objects differ from robots is that the objects are not electricity-driven or controlled by the back-end chip programming. Instead, they depend on the models pre-designed with parameters.

The current 4D printing is still in its infancy. With the rapid development of technology, it may not take long for many scenes in science fiction movies to truly happen in our daily lives through the utilization of 4D printing. Just like how we react to smart wearables, the arrival of 4D printing may come straight to us overnight when we are unprepared. What 4D printing is going to change is not only the way of assembly but also self-adaptiveness. In the future, the items we use will become super items, like a coffee cup that can adjust according to the temperature of the coffee, a sofa that can change its size in line with the different sizes of guests, and a car that is able to self-transform into an aircraft when driving into a lake to help us out of danger.

3

LIFESTYLE

The transformation triggered by 4D printing first manifests in every aspect of our daily lives, including our clothing, food, living, and transportation. From now on, food, clothes, jewelry, and even household items can be 4D printed. Unveiling the mystery of 4D printing, the world is set to become more colorful.

SECTION 1 | 4D PRINTING AT THE DINING TABLE

Currently, 4D printing has found certain applications in the food industry.

In the realm of food, researchers typically use ingredients such as soy protein, potatoes, gelatin, pumpkin, starch, and natural fibrous substances. These are designed and combined according to specific formulas and structures. Under certain environmental stimuli—such as pressure, temperature, wind, concentration differences, water, pH values, or light—the shape, properties, and functions of the printed items change over time, producing 4D printed food.

Different stimuli have varying effects on the printing materials. For example, fish paste under drying and microwave stimuli shows that microwave can cause faster gelation of fish paste, making the shape and structure of the printed product superior to that produced under drying stimulus. Therefore, for 4D printed food, the optimal stimulation method can be chosen based on different characteristics of the printing materials, such as water content, texture, and texture, to improve the quality of the printed food after stimulation.

As the application of 4D printing technology in the food sector continues to evolve, more and more food materials suitable for printing technology are being developed. Currently, materials that can be used for 4D printing food mainly include chocolate, flour, fruit and vegetable mixtures, soy protein isolate, and gelatinous ingredients, which can be combined with various cooking techniques to create delicious dishes. However, since 4D printing food technology is still in the research stage, it has not yet achieved commercial production on a large scale and is only applied in some culinary kitchens. Among these, gelatinous ingredients have a broader application in 4D printed food.

Moreover, as meat plays a vital part in our daily diet, providing essential nutrients and energy, it can also be one of the main materials for 4D printed food. However, meat is inherently difficult to print. Thus, how to better print meat materials is becoming a hot research topic in the field of 4D printing in the food industry.

Since 1960, the global population has doubled, and human consumption of animal products has increased fivefold—a number that is expected to continue growing. With countries like India becoming wealthier, many people who primarily consumed a plant-based diet are now turning toward a meat-heavy, US-style diet. However, meat production places a significant environmental burden on the planet.

As early as 2006, the United Nations released a detailed report on the crisis of global warming and the impact of human dietary habits, especially meat consumption, on climate change. In fact, climate change isn't just caused by emissions from cars and industries; the food system and dietary choices also play a crucial role in environmental degradation and climate change.

The global food system, encompassing everything from production to consumption and waste, emits a substantial amount of greenhouse gases, affecting climate and environmental change. Livestock farming, in particular, requires vast amounts of land, feed, and water resources and is a major source of greenhouse gases. According to BeyondMeat, animal husbandry produces over 50% more greenhouse gases during the rearing process.

A report published by the Standing Committee on Nutrition of the United Nations System in August 2017, titled Advocating Sustainable Diets to Promote Human Health and the Health of the Planet, predicted that global per capita dietary-related greenhouse gas emissions from crop and livestock production would increase by 32% between 2009 and 2050 if global dietary patterns shift toward higher animal protein consumption as incomes rise.

For example, producing beef requires eight times more water and over 160 times more land resources than vegetables and grains. This is because the food chain for beef production is much longer than that for vegetables and grains. Essentially, humans grow grains, which are turned into feed to rear livestock, which humans then consume. A cow needs to consume about six kilograms of grain to produce just 500 grams of meat. In the US, livestock consumes 60% of the grain. Livestock farming is also incredibly water-intensive, with 500 grams of beef requiring approximately 7,041 liters of water.

At the same time, earth's water resources are facing a severe crisis. Over-extraction of water for irrigation is causing water crises around the world. Worse still, meat production is closely linked to climate change. Livestock farming accounts for 14.5% of all human-induced greenhouse gas emissions—nearly equivalent to the emissions from all forms of transportation, including cars, trucks, ships, and planes combined.

It's this set of irreconcilable contradictions that necessitates technological solutions to humanity's meat consumption problem, with 4D printed meat being one of the key technological approaches.

Therefore, compared with genetically modified foods, artificial beef may be a better way to solve food shortages. The most important difference between artificial meat made by 4D printing and by 3D printing is that by 4D printing, it brings more convenience into our daily lives. For instance, we will no longer need to perform slicing work. By presetting the self-classifying deformation "intelligence" in the printed model, when needed, just rinse it in water, and the block of beef will automatically separate into slices (Figure 3-1).

Figure 3-1 Self-transformable beef

Furthermore, in the realm of personalized food customization, consumers can use a combination of 4D printing technology and digital cooking techniques to create distinctive 4D printed products tailored to individual needs. This approach allows for dynamic changes in the food based on user preferences, transforming dining into not just a feast for the taste buds but also a visual spectacle. This is especially pertinent in the domain of children's food, such as in family-friendly parks like Disney, where food made with 4D printing can undergo "magical" transformations triggered by specific mediums. For example, pouring water over a burger could transform it into the shape of Mickey Mouse. Clearly, 4D printing introduces a level of enchantment to food production that surpasses 3D printed foods, fundamentally revolutionizing culinary creativity.

SECTION 2 | 4D PRINTING IN FASHION

LEATHER BAGS

It is easier to apply printing technology to leather than meat products. Mainly from a technical point of view, the structure of the skin is simpler than muscles and therefore easier to produce.

From the biological point of view, the organization of meat is much more complicated than the manufacturing of fur. The fur is closer to 2D, but production of meat requires a more complex 3D technology. In addition to technology, the supervision of cultivating leather is relatively simple. A lot of products can be designed after artificial leather is produced, such as belts, leather bags, and leather clothes, among others. With the help of 4D printing, the texture of the leather will self-transform according to the different social needs and occasions of customers. Therefore, the decorative leather products will be full of variety.

In the near future, women will not have to prepare too many bags for different social occasions. All they will need is a 4D printed bag that may change into many varieties to meet different social occasions. Especially for some women who like fashion and are in pursuit of constant change, a 4D printed bag can be changed according to the needs of different styles. It can better meet the needs of women's fashion. Besides, compared to the widespread application of artificial meat, the main reason that 4D printing of leather will be easier lies in its relatively loose regulatory conditions.

FASHION AND JEWELRY

The fashion industry has a long history, especially in womenswear, of which there's never been a shortage in the consumer market—it comes in bulk, customized, with various sizes and styles. Yet, the only problem that cannot be solved is making clothes that fit all curves.

The short history of 4D printing technology does not affect its development. The US Massachusetts technological design studio Nervous System developed an elastic fit fabric and printed it into the world's first "4D dress" by using 4D printing technology (Figure 3-2).

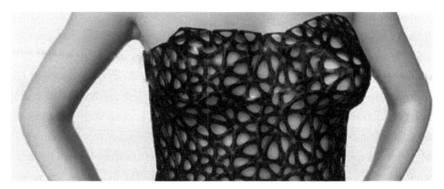

Figure 3-2 4D printed dress

The dress solves the problem of ill-fitting clothing, as it changes itself according to the wearer's body shape, and, more amazingly, it changes its shape automatically. This is perhaps the dress that is missing from every woman's wardrobe. The fabric fibers used to make the dress are made of 2,279 triangles linked by 3,316 built-in hinges. The tension between the triangles and the hinges varies with the shape of the human body, so no matter if you gain weight or lose it, this 4D dress will never fail to fit.

The space taken up by the dress inside the printer is much bigger than printing a regular 3D model dress, but technicians still managed to print it out in one piece. Meanwhile, researchers also 4D printed a series of jewelry to match the dress, which undoubtedly could also change automatically based on the shape of the human body.

The principle of 4D printing material used by the Nervous System was a breakthrough based on MIT's model. It applies "Selective Laser Sintering Technology," which uses a laser to sinter powdered nylon material to create prototypes. The powdered materials between the triangles and the linked hinges are not sintered by the laser and drop after printing, and thus, interlocking fibers are created. It takes 48 hours to produce each dress, of which the cost is £1,900 (about RMB 18,393). However, the price does not discourage women.

The 4D printed dress means that the shape of a 3D printed object can transform automatically into another shape, allowing the object to transform into its final design form without manual labor. Nervous System has developed an application that enables users to 3D scan their body, then select a fabric size and shape, and instantly customize a unique 4D dress.

A long time ago, we were able to print parts through 3D printing and then manually put them together to create a large object. However, what makes 4D printing different is that the printed object can be automatically assembled or transformed into a predetermined shape. In contrast to MIT's idea, they named this process "Kinematics" after the branch of mechanics of the same name—also known as the geometry of motion, which describes the movement of an object rather than its cause.

Rosenkrantz said, "We think the greatest advantage of 'Kinematics' is that it can transform any three-dimensional shape into a flexible structure for 3D printing. The system then compresses the structure down through computational folding."

To make these dresses, a 3D scan of a person's body form is essential and is the basis of the digitally modeled garment. The hardness and the state of the final product can be controlled at this stage by selecting the structure of the triangular hinged mesh, and the way the material drapes can also be simulated on-screen. With computer simulation software, this digital model can be folded into a much smaller shape and then printed out in a compressed form. When the dress is lifted out of the printer, it will automatically unfurl into its predetermined shape.

According to Rosenkrantz, "Compressed designs offer benefits not only for production but also for transport. It holds great promise for the creation of flexible wearables but could also be used to enable the production of other large scale structures in today's small-scale printers." Before then, in response to the request from Motorola, Nervous System began developing the concept of Kinematics to customize 4D printed products.

Kinematics can generate a design composed of 10 to 1,000 different parts which can be expressed to one another to build a dynamic or mechanical structure. When printing a dress, for example, first, one must scan a user in

3D to sketch the dress, and then inlay a pattern on the dress, followed by generating the Kinematics structure to simulate the draping features of the dress. Finally, the dress which is much bigger than the printer, is compressed and then computationally folded. The dress can be printed in one piece instead of different parts and combined into one.

The price of these products varies depending on the customization options. As long as customers are happy with their customized design, after placing an order, Nervous System will print it out. The second free application allows users to experiment with the Nervous System's templates and print the results at home. Rosenkrantz and Louis-Rosenberg further developed their theory by adding the ability to fold the design into its smallest spatial structure.

As early as 2015, 4D printing technology was already being explored for use in apparel. Karl Lagerfeld showcased Chanel's iconic classic suits made with 3D printing technology at the Chanel Fall/Winter 2015 Haute Couture Show, dazzling the fashion world during Paris Fashion Week. However, the comfort and difficulty in modification were significant drawbacks of 3D printed clothing. Now, these issues have been addressed with the advent of 4D printing technology.

As Karl Lagerfeld said, "The vitality of fashion lies in its ability to keep pace with the times. The fashion industry needs to integrate with new technologies to design clothes that customers prefer and are more willing to accept."

A design studio in the US, Nervous System, previously designed a stunning dress that was 3D printed by Shapeways and has been acquired by the Museum of Modern Art in New York. The dress, designed by Jessica Rosenkrantz and Jesse Louis-Rosenberg, utilized Nervous System's 4D printing system, Kinematics, to create complex, foldable shapes from a single piece, making it fully wearable. Thousands of "fabric pieces" connected by hinges fold smoothly and conform to the body.

The most remarkable feature of this garment is its ability to adapt to the environment, fitting the wearer's body movements perfectly. Theoretically, a mature 4D printed dress could serve as a lifetime garment for someone, automatically adjusting as the person grows, changing styles according to social occasions, and adapting to the climate.

Beyond these explorations in 4D printed apparel, today, many stimulus-responsive 4D textiles have also made breakthroughs. For example, ChroMorphous fabric (Figure 3-3), which changes color in response to stimuli, represents an innovation in the textile industry. This user-controlled color-changing textile allows users to control the color and pattern of their clothing and accessories through a smartphone.

Figure 3-3 ChroMorphous fabric

ChroMorphous was developed by scientists at the University of Central Florida specializing in fibers and photonic devices. Unlike past color-changing fabrics that required sunlight or body heat to work, ChroMorphous transcends the limitations of other color-changing garments and textiles by offering on-demand color change.

Each fiber in ChroMorphous contains a conductive micro-wire. When an electric current passes through the micro-wire, the fiber slightly heats up, activating its color-changing pigments to produce various colors and patterns. The subtle temperature change produced during the fabric's color change is almost imperceptible. Like traditional fabrics, ChroMorphous can be cut, sewn, washed, and ironed. Imagine a backpack or other accessory that can change color to match your outfit for the day or walls and curtains at home that change color with the seasons or your mood. Although currently, there are only four colors to choose from, it's enough to captivate anyone's interest.

The advantages of 4D printing are, on the one hand, the ability to compress designs into their smallest spatial structures and then 3D print them so that the printed objects exclude redundancy. On the other hand, the printed objects can change to fit different needs. All these benefits are unmatched by 3D printing.

The current operating pattern of the fashion field will be changed through the implementation of 4D printing. For example, no clothes have to be displayed, and no inventory is needed. Instead, a human body 3D scanner, a computer with professional design templates, and a virtual reality mirror will be needed. Users could then choose the template from the computer-based choices based on their style and preferences, and the template could be of any style or even their photo. Then, the system will automatically design and generate the corresponding garment.

The model, auto-generated precisely by the system with the help of a human 3D scanner, will scan the size of the human body and then merge it with the design. Customers will then be able to see a simulated display that is close to the real. Through the virtual reality mirror, customers could rotate the model for display at any angle and adjust and modify it as they wish. Finally, the garment that they had in mind would be generated. After that, the virtual garment will be printed out by transmitting it to the printer via the cloud service. Alternatively, if the store has a 4D printer, the garment could be printed out directly in the store. As such, 4D printing will bring fashion into the era of private customization.

FASHIONABLE HIGH HEELS

It is not news that 3D printing technology has made its way into the fashion industry. A collection of 3D printed women's high-heeled shoes (Figure 3-4) was exhibited at a futuristic fashion show at the World Fashion Center in Amsterdam. It comes from a graduation project of Pauline Van Dongen, who is a graduate of a Master's in fashion at ArtEZ University of the Arts in Arnhem, Netherlands.

As seen from the figure below, there is one thing that is common with 3D printed shoes—they are seamless overall. The mesh design breaks the traditional impression of shoes and feels nondurable. However, that's not the case. The shoes are actually printed from nylon, which is not only lightweight but is also strong and tough. What's more, their leather insole is patented, and the entire outer layer of the shoes is coated with synthetic rubber, which can increase friction, improve durability, and provide comfort of wear.

Figure 3-4 3D printed high heel shoes

At the same time, people in related industries from China are actively involved in not only printing villas, but fashionable shoes, as well. The entrepreneurs in Tianjin printed fashionable shoes by using 3D printing technology (Figure 3-5).

Figure 3-5 3D printed shoes in China

Neither 3D printed shoes nor traditional shoes can solve one problem, that is, the independent variable of shoes. In other words, the shape and size of the shoes we buy are fixed and will not change according to the thickness of the socks we wear, nor will they change their shape and size while we are climbing or walking.

With 4D printing, however, this is a different story. 4D printed shoes will thoroughly change our definition and perception of shoes.

A pair of 4D printed shoes (Figure 3-6) from MIT's self-assembly lab, formed entirely on their own, and could even contract around the feet once they are put on. According to Skylar Tibbits, the leader of 4D printing in the US, 4D printed shoes are self-forming and self-contractible, and the contractible materials used can minimize the cost of making shoes. Additionally, the materials used in 4D printing are even more "magical" as they can change their shape in response to external stimuli, such as temperature, water, pressure, and so on.

Figure 3-6 4D printed shoes

With such "capricious" shoes, in the future, we don't have to worry about not being able to get the correct shoes. Regardless of size or style, the future 4D printed shoes can fulfill our customization needs. After users try on their favorite shoes in the smart mirror, the 4D printer will automatically print out the corresponding shoes according to the users' foot size and bone density. This will solve all "fitting" problems.

The shoes of the future will self-adjust their size and shape according to not only the swelling of our feet and the thickness of the socks we wear but also the different occasions. At least some changes are foreseeable. First, women will no longer have to be embarrassed when their boss or friend suggests mountain climbing. The high-heeled shoes on their feet will automatically change into flat shoes that are suitable for mountain climbing or long-distance walking. Second, we don't have to worry about the different sizes of shoes required in different environments, nor do we have to worry about the looseness of shoes long-term wear because 4D printed shoes will automatically change and adjust to maintain the appropriate size for us. Even in the future, the full intelligence of 4D printed shoes will be realized in their applications, like switching to ankle protection mode while playing basketball, switching to grass walking mode while walking on the grass, and switching to waterproof mode while walking in the rain.

Problems like shoe' size, fit, and so on are only the first hurdles solved by intelligent technology and 4D printing in the field of footwear and apparel. In the future development, with 4D printing technology, more applications will bring human beings magical use and enjoyment. Nothing is impossible!

SECTION 3 | THE REVOLUTION OF UNDERWEAR WITH 4D PRINTING

There are significant developmental changes during the different stages of female breast tissues. Even in adulthood, when breast development has matured, many women have breast fullness, stiffness, and tenderness before menstruation. In severe cases, the slightest movement can be very painful and uncomfortable. This is due to the increased estrogen in the body before menstruation which leads to breast hyperplasia and edema.

The current female underwear is similar to traditional clothing and shoes. Its size is fixed and will not change according to the changes in breasts. Some researchers used uniform 3D printing to print underwear (Figure 3-7). The unusual aspect of the bikini is that it is printed by a 3D printer. The outline of the underwear is "woven" by circular patches connected to each other, and the thinnest part is only 0.7 mm. By using nylon 12, its texture is very soft and not easy to break. According to the researchers, this underwear is very refreshing and comfortable to wear after it is soaked, and it may be more suitable for swimming.

As the 4D printed underwear, compared to the 3D ones, can self-transform and self-assemble, it will change the current fact that the size and shape of the underwear cannot adapt to the changes in breasts. Comfortable underwear-wearing experience will be available through the 4D printed underwear that can self-transform according to the changes in breasts before and after menstruation. This may completely change the entire female underwear market.

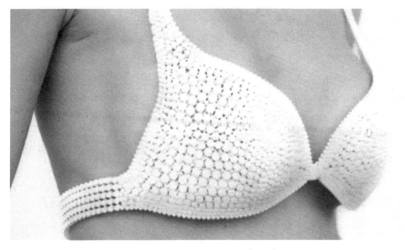

Figure 3-7 3D printed underwear

SECTION 4 | 4D PRINTING AND SEX TOYS

If both 3D and 4D printing want to enter the sex toy market, the following problems need to be addressed.

Safety and health. For example, the current 3D printed products, especially those with low-end molding, face low-precision problems and are often very rough on the surface. In the field of sex products, users will hardly be able to get satisfaction with such products. Besides, the rough texture can easily breed bacteria. These problems can't be ignored in 4D printing.

Privacy. In terms of the current climate, it takes quite some time for 3D or 4D printing devices to enter a household. Users still need a relevant service provider to process their personalized sex products, which involves questions about the user's privacy. This may become a considerable concern while users are selecting their service provider.

Materials. The materials used for sex toys are mainly silicone, particularly used for the parts that have direct contact with sexual organs. The current materials used for either 3D or 4D printing are unable to satisfy the actual usability of sex products, especially the key technology, "smart" materials, of 4D printing. But, as technology advances, the problem of materials will be solved.

As such, if 4D printing cannot reach the household application level, even in Europe, the US, and Japan, where sexual concepts are relatively open, people will not be willing to discuss sex products in public and will also be embarrassed about receiving them. After all, it is a private matter.

On the other hand, just as there are no two identical leaves in the world, each person's physiological structure is also different. So, the sex products already in the market may not be suitable for all, which is why 3D printing is currently popular.

The emergence of 4D printing technology makes it easier to privately customize sex products. The products can be designed and printed based on the individual's physiological structure. In addition, a 4D printed sex product can self-change while being used along with the user's physiological response during the different stages to facilitate an optimal experience.

When the design technology is further developed, a human body model can be generated and designed with only a photo. The most significant difference between the 4D printed human body model and the inflatable doll is that the model can be designed and printed with the image of a user's dream lover. Meanwhile, it would also self-change with the different physiological responses of the user during use. No matter what cup size you want to experience, A cup, B cup, or D cup, the 4D printed sex toy can self-change to bring you comfort and pleasure.

Similarly, for female users, regardless of what kind of smart "egg," smart "stick," or 3D customized toy is used, they are hardly able to make adaptive changes to the different physiological responses of the human body. Hence, for the sexual organs that are sensitive and delicate, it is difficult to provide the maximum "intimate" care.

The situation is entirely different with 4D printing technology. It doesn't matter if it is an "egg" or a "stick," as long as the trigger is set while printing, which can be the secretion of sexual organs, a kind of breathing rhythm, a sound or an intensity, among others, this magical item will self-change its length or size with the changes of user's physiological response.

If we add a 4D printed sex product to the robot in the future, do we still need real partners? Without a doubt, when 4D printing is implemented intensively in the field of sex products, a new ethical discussion will be inevitable.

SECTION 5 | 4D PRINTING MEETS SUPER DADS

As a new dad, it is a joy to spend time with your baby. But when it comes to making formula, changing diapers, undressing the baby before bed, and lively telling bedtime stories … these tasks are headaches to most dads. Although the advent of a formula dispenser machine that can instantly make a bottle of warm formula has been a solution to one of these headaches, the rest are still bothering them.

4D PRINTED DIAPERS

For example, diapers, for most new dads, make the babies upset because of their short-lived patience. It is much easier to clean up a diaper than to put on a new one. It usually takes several adjustments to put on a diaper properly, and the entire process is difficult for babies who cannot focus and cooperate. If the diapers are manufactured using 4D printing technology, as long as the position of the buckle does not greatly deviate from product to product, the degree of tightness will be self-changing and self-adjusting to achieve optimal control (Figure 3-8). On the other hand, when the capacity of a diaper exceeds the normal, and there is no time for a change, it will be able to self-change according to the humidity as the triggering medium. It will expand the capacity to accommodate the excess.

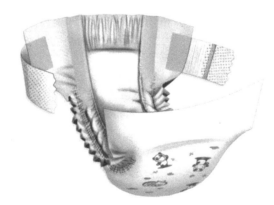

Figure 3-8 Diaper

It's clear that this technology is also very important for women, as they experience menstrual periods every month. During the menstrual cycle, the amount of flow is not always constant and can fluctuate based on the body's condition and age. Facing these fluctuations, using 4D printed sanitary pads or sanitary pants, which can automatically adjust their capacity based on the volume of blood as a trigger medium, can effectively address women's concerns and make life more humane thanks to technology.

4D PRINTED BABIES PAJAMAS

Another example is that babies don't want to sleep undressed, but they can fall asleep while wearing clothes. When sleep and the changing of clothes come together, perhaps this can also trouble a father. However, if the clothes are manufactured using 4D printing technology, it could be very different.

On the one hand, when a baby falls asleep, the 4D printed clothes worn by the baby will self-transform into a quilt; on the other hand, in line with the weather and the thickness of the clothes, they can self-assemble and self-transform to reach the best conditions. Of course, it's also possible to adjust clothing based on a child's sleep state. For example, before the child has fallen into deep sleep, the clothes can be appropriately loose to accommodate the child's movements. Once the child has entered deep sleep, the clothes will automatically tighten to better provide warmth.

SECTION 6 | 4D PRINTING ENABLES TOYS TO TRANSFORM AUTOMATICALLY

With the continuous improvement and enhancement of material life, the market demand for the toy industry is continuously expanding, especially the modern toy industry. Modern toys in China started in the mid-to-late 1980s when the industry emerged accompanied by the Reform and Opening Up. The toy industry is labor-intensive in the manufacturing sector; it is the same as the textile, clothing, shoes, and plastic products industries. The so-called educational toys nowadays, also fall under the category of traditional manufacturing.

According to publicly available industry information, China, as a major toy manufacturing country, has a large number of toy manufacturing enterprises. Over 70% of the world's toy products are produced in China. From 2016 to 2021, China's toy export value continued to grow, from US$18.39 billion in 2016 to US$46.12 billion in 2021. Among them, the US is a major export market for Chinese toys, with exports to the US amounting to US$13.8 billion in 2021, an increase of 57.3% over the previous year, accounting for 29.2% of China's total toy exports. In 2022, the market size of China's toy industry reached RMB 91.8 billion, and it is estimated that the market size of China's toy industry will reach RMB 119.5 billion by 2026.

Although this looks good as data, from the perspective of the overall independent design and creativity of the toy industry, China's toy industry is still in its infancy. It's not unusual for a country that is mainly at the stage of industrial manufacturing to ignore the cultural and creative elements of the toy industry.

Influenced by factors like the increase in national income, the pressure of the economic downward trend, and the national support for industrial transformation and upgrading, as well as the cultural creative industry and Makers' activities, it is inevitable for the toy industry to develop rapidly. Such development must be closely related to 4D printing.

The toy industry is not limited to the traditional concept of children's toys. Toys should be thought-leading tools that bring people of different age groups enjoyment and help them explore their lives. This will inevitably lead to niche products and demand for creative and personalized products. Obviously, traditional large-scale manufacturing fails to satisfy the new development of the toy industry. The small batch, personalized, and rapid manufacturing will become the critical production technique that will achieve the rapid growth of the toy industry. The rise of 3D printing technology is, to a certain extent, an excellent solution to this problem.

The application of 4D printing technology may not only change the in-bulk selling ideas of the traditional toy industry but also bring the toy industry into the field of creativity. For instance, children's needs for toys keep changing, especially for those under the age of five, whose curiosity and exploratory skills are quite strong but relatively weak in concentration. To satisfy their needs, some families may bring their children to professional children's playgrounds to fulfill their curiosity, while some families keep buying their children new toys.

However, such methods have brought trouble to parents, like the inefficient placement of toys in an expensive house and difficulties in meeting the changing needs of children regardless of any methods. Perhaps 3D printing technology can be a good choice; that is, each family can be equipped with a 3D printer, and every day, parents could print out toys from the printer with the materials and models purchased. This may be a good solution, but regarding 4D printing technology, it is obviously not suitable.

Through 4D printing, when printed toys are not needed, the materials can be recycled and reprinted many times. Alternatively, printed toys can exist without a fixed form. Just like the popular toy nowadays, Transformers. The only difference is that the 4D printed Transformers could also self-assemble and self-transform rather than be assembled or disassembled by children.

This is disruptive to the toy industry for children between 0 and 3 years old. The Makers are the people who benefit the most from the concrete changes brought about in the era of 4D printing technology. Makers are a group of people who share the same idea of pursuing innovations, and the toy industry is a challenging creative industry, especially so in adult toys, which have challenges in innovative thinking. For the Makers, on the one hand, there are many new market opportunities, and it is an industry worthy of challenges and exploration. On the other hand, there are new challenges to the children's toy market, like how one can design a toy that can self-transform into a duckling, a chick, or a puppy. This is the innovation of the toy industry driven by 4D printing technology, which has far-reaching significance for the education of both, parents and their children (Figure 3-9).

Besides, for traditional toy manufacturing companies, in the era of 4D printing, the business model will no longer be about selling toys, but instead, it will be about selling innovations or providing 4D printing services. Consumers could either purchase the existing creative toy models directly through the Internet platform or put forward their innovative ideas of toys they want to make. Those would then be processed online by professionals like Makers or toy designers, and then printed out on the designed model either by the designer or by the nearest printing service provider appointed by the consumer. Of course, consumers would also be allowed to have the purchased models printed out by their own 4D printer.

Figure 3-9 Self-transforming toys

To put it briefly, in the era of 4D printing technology, the future of the toy industry is no longer in selling toys, but in creativity. Of course, the sale is not about the materials, as the materials can be recycled, which not only solves the problem of energy-saving and environmental protection but also minimizes the consumption costs for consumers.

Most importantly, the 4D printed toys that can self-assemble and self-transform with given instructions can not only satisfy children's curiosity and stimulate their creative thinking but also help parents better meet the needs of their children.

For the adult toy market, 4D printing technology will pose a challenge for Makers and will undoubtedly inspire the industry's development. Our toys may take us into a toy mobile maze, which will have profound significance in our thinking and exploration. Likewise, 4D printing will also change the training industry, as well as the assessment of HR recruitment. Maybe soon, the interviewer will no longer give a paper-based test but a ready-to-change 4D printed toy, like a mobile maze, where the candidate will need to find the variables and complete the toy within a specified period (Figure 3-10).

Figure 3-10 Toy with variables

SECTION 7 | 4D PRINTING IN HOME FURNISHINGS

Most people have probably been in similar situations where they were annoyed because of the lack of tools at home, such as scissors, wrenches, and other simple tools. We can print them out with the help of 4D printing technology, and what differentiates it from 3D printing technology (Figure 3-11) is that instead of having to make a single print for each tool, we can design the variations of multiple tools directly into the printing model. Then, we can trigger the medium to self-transform and self-assemble the printed tool.

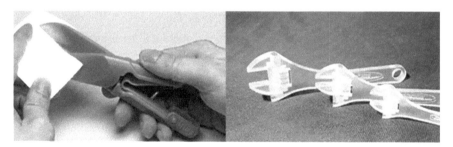

Figure 3-11 3D printed tools

Even the lack of a stool, chair, ladder, etc., is not a problem. Despite this fact, 3D printing technology is also sufficient for printing these objects (Figure 3-12), and 4D printing technology is incomparable in this aspect. For example, when we need a ladder, with 3D printing technology, we must print it out at a predetermined height, and when we do not need it, it will inevitably take up space. But with 4D printing technology, the ladder can be designed in an optimal and minimal form, and the height that it can extend could be set in the printing model so that the ladder could be triggered to self-adjust to different heights to fulfill different needs under various environments. It can also return to its "original form," making it a stool or chair for everyday use when the user does not need it.

Figure 3-12 3D printed chairs

As for kitchenware such as spoons, cups, plates, pots, bowls, scoops, and others, they are not a problem for 3D printing technology (Figure 3-13), as well. However, these 4D printed objects are far beyond their 3D printed cousins. There is one significant feature of 4D printed kitchenware, that is that it can self-transform based on the amount of food cooked. We don't have to worry about the situation whereby the pot is too small to cook sufficient food to serve our guests.

Figure 3-13 Kitchenware

SECTION 8 | 4D PRINTING REVOLUTIONIZES HOUSING

The 3D printing technology has been around for quite some time, and it's only been several years since the capital boosted it and made it popular. Pushed by capital and talent, many geeks, researchers, and entrepreneurs have made various attempts based on 3D printing technology, including printing techniques, printing equipment, printing materials, and so forth. They have also explored 3D printing in different fields, such as printing artwork with sand, printing circuits with graphite, and printing hearts or limbs. What is unique is that they used 3D printing technology to print houses, which could probably revolutionize the construction industry.

Actually, 3D printing technology in the construction field and the printing of houses is not new. However, it has indeed brought revolutionary techniques to the construction industry. Its advantages can be mainly reflected in the below aspects.

Energy-saving, environmental protection, and the saving of materials. This is because there is no dust or construction waste generated in 3D printing buildings, which is good news for city dwellers who are suffering from the haze and construction noise.

Turn waste into treasure and realize the secondary use of construction waste. All kinds of construction waste, industrial waste, and mine tailings can be used as raw materials for 3D printing. In particular, the sand of the desert is like "gold" for 3D printing—it is an excellent raw material. It can be used to make sand fixing walls and vertical greening walls, among other things, as well as realize sand fixation and help to control the desert.

Save construction costs and improve construction efficiency. In regard to the 3D printed buildings we have actualized so far, the application of the technology can save 30% to 60% of building materials, shorten the construction period by 50% to 70%, and save labor by 50% to 80%. That means, with 3D printing technology, it is possible to build a house at a lower cost and within a shorter period, saving at least 50% on construction costs.

In its practical applications, the concerned 3D printed buildings are in Suzhou. These include a villa with an area of about 1,100 square meters, a five-story residential building, and a simple exhibition hall, of which the walls were 3D printed using "ink" composed of a little steel, concrete, and construction waste (Figure 3-14).

Figure 3-14 3D printed building

3D Printing: How Are the Walls Printed?

In the light of the Suzhou 3D printed buildings, the "printer" used was 6.6 meters high, 10 meters wide, and 32 meters long. Its foundation was as broad as a basketball court, and it was three stories in height. The width of the printed material can only reach 1.2 meters, but the length can be unlimited. According to the computerized design drawings and plans, a large nozzle controlled by the computer ejected "ink," similar to cream decoration, and the "ink" was stacked in a Z shape layer by layer. Soon, a high wall was built. Afterward, the walls were built up like blocks, then joined together by infusing with the reinforced concrete in the second "printing." In this way, ten buildings of 200 square meters each could be printed out within 24 hours.

WILL 3D PRINTED HOUSES BRING DOWN HOUSE PRICES?

Given the current market price of an 80-million villa, the cost of 3D printing is just over one million. Other than the advantageous price, in terms of efficient construction, it is way beyond traditional construction, and such a villa can be printed out in a day.

ARE 3D PRINTED HOUSES SAFE?

Although there are no specific monitoring standards for 3D printed buildings, or relevant standards for 3D printing in the construction industry today, from the perspective of developers' tests, 3D printed walls are five times stronger than the ones printed with ordinary cement. With the help of the 3D printed walls, a more complex and solid structure can be achieved, and the hollow wall structure can create a living environment that is warm in the winter and cool in the summer.

To obtain the large-scale commercialization of this emerging technology and its applications, the overall performance of stiffness, strength, and durability, among other factors, has yet to be further validated. Also, relevant testing standards are yet to come.

CURRENT APPLICATIONS OF 3D PRINTED ARCHITECTURE

Japanese 3D printed architecture. Serendix designed a small house with 3D printed walls for less than US$25,500, completing the project in less than 24 hours. Although it's hard to imagine living in such a house, this project showcases the flexibility of 3D concrete printing. The building's surface area is only ten square meters, with a honeycomb structure and no reinforcements. Serendix aims to construct emergency housing during crises with frequent earthquakes and typhoons. Assembling different 3D printed parts takes three hours, with a total of 23 hours and 12 minutes required to complete the entire building. (Figure 3-15)

Figure 3-15 Japanese 3D printed architecture

Virginia's architecture in the US. Alquist 3D, a company based in the US, previously announced its project of 3D printed houses. The project involves building 200 houses in Virginia to reduce the overall infrastructure cost in economically troubled communities. According to the company, the plan aims to be one of the largest housing construction projects, with Pulaski and Roanoke being the first selected cities. They plan to further expand into the 3D building market after completing the Habitat for Humanity project.

Canadian 3D printed architecture. Fibonacci House is the first fully 3D printed house listed on Airbnb. Printed by Twente Additive Manufacturing and created following the famous Fibonacci sequence, its design is intriguing. The house adopts a spiral shape, allowing space to extend from the exterior to the most compact part's enclosed and warm space. This tiny house can be rented on Airbnb in Canada for at least two nights at a rate of US$128 per night. (Figure 3-16)

African 3D printed architecture. Mvule Gardens is Africa's largest 3D printed house project, consisting of 52 houses. The project, built by Holcim and CDC Group's joint venture 14Trees, aims to address Kenya's housing shortage. These houses are constructed using COBOD's BOD2 printer and

Figure 3-16 Canadian 3D printed architecture

Holcim's TectorPrint dry mortar, ensuring the houses are sturdy and reducing the carbon footprint. 14Trees offers two-bedroom and three-bedroom houses, hoping to help Africa achieve green, low-income, and affordable housing. (Figure 3-17)

Figure 3-17 African 3D printed architecture

3D printed architecture in California, US. To develop housing more quickly, economically, and with less environmental impact, Ross Maguire and Gene Eidelman introduced Azure 3D Printed Homes in 2019. Utilizing years of experience in building and developing properties worldwide, Azure 3D Printed Homes can print eco-friendly houses within 24 hours using recycled waste. The company offers various backyard offices and residences, including the Azure Marina model. (Figure 3-18)

Figure 3-18 California's 3D printed architecture

3D printed architecture in Virginia, US. The nonprofit organization Habitat for Humanity uses 3D printing technology to help low-income families solve housing problems. Last year, in collaboration with Alquist 3D, they built the first owner-occupied 3D printed house in Williamsburg, Virginia. The house features a kitchen, three bedrooms, and two fully equipped bathrooms. Habitat for Humanity plans to build 3D printed houses in other states as well. (Figure 3-19)

European 3D printed architecture. House 1.0 is Europe's first small 3D printed house, manufactured by Denmark's 3DCPGroup using a concrete 3D printer. The house's design aims for a better, faster, and more environmentally friendly construction process, reducing labor. Despite its 37 square meters area, the small house provides all the necessary facilities and is cost-effective. (Figure 3-20)

Figure 3-19 Virginia's 3D printed architecture

Figure 3-20 European 3D printed architecture

Czech 3D printed architecture. Prvok is a 3D printed house in the Czech Republic, created by the Buinka company and sculptor Michael Trpák. It can be placed in rural, urban, or even aquatic settings. Constructed of concrete at a printing speed of 15 cm per second, it only takes 22 hours to complete, measuring 13.35 meters × 3.5 meters × 3.1 meters. A robotic arm can create a living area of 43 square meters. The house includes a bathroom with a toilet, a living room with a kitchen, and a bedroom. Additionally, the building can be mounted on a floating base, allowing for aquatic living. (Figure 3-21)

Figure 3-21 Czech 3D printed architecture

French 3D printed architecture. Viliaprint is a construction project that combines additive manufacturing with traditional building techniques. Launched in June this year in the French city of Reims, located in the ecological district of Réma'Vert, it involved the collaboration of stakeholders to construct five houses, with living spaces ranging from 77 to 108 square meters. This promising project aims to improve social, economic, and sustainable development aspects. The 3D printing of the houses was not conducted directly on-site, and the designers emphasized the rapid drying of concrete after each layer's application to support its weight. (Figure 3-22)

Figure 3-22 French 3D printed architecture

German 3D Printed Architecture. PERI, established in 1969, acquired a minority stake in the 3D design leading company COBOD in 2018 and has cooperated ever since. They built Germany's first 3D printed house and Europe's largest apartment building. Located in the Rhineland region of North Rhine-Westphalia, Germany, this two-story single-family home covers 160 square meters and features three layers of insulated hollow walls. The construction used the COBOD BOD2 printer, characterized by extruding concrete at a speed of 100 cm per second, achieving the desired project size, smooth and straight walls, and the highest quality requirements. Using the COBOD printer allows constructing up to 300 square meters of space on three floors at unprecedented speeds. (Figure 3-23)

Figure 3-23 German 3D printed architecture

3D printed architecture in the Netherlands. The next housing project is located in Eindhoven, Netherlands, consisting of five 3D printed concrete houses. Though these houses have been sold, they can be rented from the real estate company. Each house is made up of 24 individual concrete blocks, requiring 120 hours to print layer by layer. These 24 parts are then moved to the site and assembled, connected to the foundation, and fitted with roofs, windows, and doors. These buildings have a futuristic design, making them appear like rocks in an oasis surrounded by trees. These houses meet the highest comfort requirements and are built in a sustainable and energy-efficient manner. The surrounding environment is bright and tranquil, making it a great place to relax. (Figure 3-24)

Figure 3-24 3D printed architecture in the Netherlands

3D printed architecture in China. In China, a team led by Professor Xu Weiguo from the School of Architecture at Tsinghua University used robotic 3D printed concrete construction technology to print several residences for farmers in Xiahuayuan Wujiazhuang, Hebei Province. These farmhouses are functional, aesthetically pleasing, structurally solid, and ecologically energy-efficient, covering a total area of 106 square meters. The design adopts the local traditional cave dwelling form, including a residence with three large and two small rooms, where the three large rooms serve as living rooms and bedrooms with barrel vault roofs, and the two smaller rooms are used for the kitchen and toilets. The printing construction used three sets of robotic arms with 3D concrete printing mobile platforms, positioned in the middle of the three large openings for on-site printing of foundations and walls. Meanwhile, the barrel vault roofs were prefabricated and assembled onto the printed walls using a crane. The exterior walls of the building are printed with a woven texture as decoration, forming an exterior wall system that integrates decoration, structure, and insulation. The entire printing platform requires only two people to operate buttons on the mobile platform to complete the entire house's printing and construction, integrating and simplifying the process of concrete 3D printing, and minimizing labor input.

From the 3D printed architecture around the world, it's almost certain that 3D printing in the construction field is increasingly accepted by more people. Not only because of higher construction efficiency but also due to lower construction costs, higher performance, and the ability to build according to the personalized design requirements of architects. More importantly, the construction is safer and no longer dependent on traditional construction workers.

What Will Be Disrupted with 4D Printing?

The appearance of 3D printing in the construction industry not only enables the customization of buildings but also effectively reduces construction costs and increases construction efficiency. This will have a revolutionary impact on the existing real estate market. But when architecture meets 4D printing, the enhancements may not only be in-house prices but also the below two aspects.

The efficient use of living space. The space we live in today seems to be an issue that affects the quality of our lives, and all of it is due to the high house prices. To think further, the size of floor space truly has an impact, but the rational utilization of space seems more dominant. The houses we live in today, let's say the living room, are often not utilized. This includes our bedrooms. Generally speaking, while using either a bedroom or living room, the other room stands idle, whereby a contradiction arises between inadequate space and inefficient utilization.

4D printing can efficiently solve this issue. When we are in the living room, the spatial pattern of the house could automatically change, merging the space of the bedroom into the living room to create extra room. And vice versa, when we need to rest in the bedroom, the spatial pattern will similarly auto-change, merging the space of the living room into the bedroom to provide a more comfortable resting area.

The impact on the renovation industry. Renovation is not needed for 4D printed architectural space because the elements of the renovation wanted by the residents would be integrated into the initial printing and printed out directly with the printing technology. Not only that, but the 4D printed architectural style, renovation style, and renovation layout can also all change depending on the residents' ideas. We can also say that the style of home renovation can change daily with our moods.

All we will need to do is to give our ideas for changes to the model design of the building at the early stages of the printing process. Then, the environment, layout, and style of the interior renovation will self-change according to the trigger instructions from the residents. A 4D printed building may no longer be a building that we live in but a "magic cube."

SECTION 9 | CARS WITH 4D PRINT

Centennial BMW

Centennial BMW released a brand-new concept car, "BMW VISION NEXT 100," that represents the future of BMW and the direction of its latest technologies. These new technologies and materials were all produced using 4D printing. Its 4D printed body parts are functional and at the same time, by programming the printed objects, they also give the car the ability to change shape in the future.

Functionality. All of the car's 4D printed parts will be functional. Today, the design, manufacturing, and functionality of the parts are all worked on separately. In the future, vehicle parts will not only be manufactured using today's additive manufacturing process but will also be completely functional after printing. We could say that the vehicle is going to be "growing" out of a pile of raw materials.

Transformability. 4D printing technology is not producing specific parts, but new materials that are smart and interconnected. The hardness and the water solubility vary depending on the arrangement of the tiny fibers within these materials. Researchers use such properties to "encode" the printed objects to enable them to transform into a more complex shape.

The innovation and application of materials will be the core of automotive manufacturing's future. Carbon fibers and new composite materials are already being used in current BMW models. The new materials used in the concept car, such as fibers made from recyclable and renewable materials, will be realized by 4D printing and rapid manufacturing in the future, while the application of these new techniques will bring earth-shattering changes to the field of automotive materials and manufacturing. Additionally, as the techniques of rapid prototyping and rapid manufacturing become more prevalent, the smart and interconnected materials produced by these techniques will make the automobile a product that is way more complex and flexible.

SAFETY PROTECTION EQUIPMENT

An example of this is the car's airbags. In the event of a car accident, they can quickly eject to protect the people in the car, minimizing the injuries suffered by the people sitting in the front due to the impact of the car crash. However, the level of protection of the airbag is still limited, and the ejection of the airbag itself has a certain level of impact on the human body.

It would be an excellent idea then if there was an object that could be transformed at the moment of a crash to protect the human body from its impact. With 4D printing and the successful development and application of new polymeric materials, the introduction of such safety protection equipment is no longer a dream. All that's left is a matter of how long we need to wait for this dream.

The 4D printed safety protection equipment, first of all, will protect essential parts of the human body much like a bulletproof vest. Further, it will also look a lot like the legendary golden shield and iron shirt* (the most famous hard *qigong* in Chinese kung fu), whereby the wearer would be able to withstand punches and kicks without any damage, and even ordinary swords and knives could not hurt them. Furthermore, the wearer could also reach the levels of strong qigong to protect the entire body, so that it can obtain unimaginable effects such as not drowning in water, not burning in a fire, not dying while holding one's breath. Besides, it will be lighter, thinner, and smaller than a bulletproof vest in its volume. It can even be internalized into the human body, just like how the golden shield and iron shirt, so that once it is subjected to external force, it will

---

* The golden shield and the iron shirt can sustain the thrusts of sharp weapons on one's bare skin.

instantly appear to protect the human body from harm. Therefore, such safety protection equipment should not only be applied to driving cars but should also be worn by people in daily life just in case of emergencies.

SECTION 10 | 4D PRINTED UAV

Nowadays, owning an aircraft toy or a drone is nothing new. From toys to civil and military use, driven by capital, Unmanned Aircraft Systems (UAVs) (drones) have developed just as rapidly as 3D printing. When the Chinese New Year of 2015 had just started, Zhang Ziyi was proposed to by her boyfriend by placing a diamond ring on a UAV flying slowly toward her. Since then, the UAV, as a technology star, crossed over to the entertainment circle.

From the perspective of UAV development, one point of difference is that the community of Makers boosted its current craze. Up until now, we could control a small helicopter with the help of a mobile application. For people who love to stay at home, it can provide logistics service to deliver food, or by pairing it up with a camera, one could experience God's point of view by exploring nature through aerial photography (Figure 3-25). One could even propose marriage.

Figure 3-25 UAV photography

The current UAVs are no longer a mystery. Many of them are controlled by mobile devices, which may not even be considered new. A 3D printed UAV, driven by Makers, has taken to the skies (Figure 3-26), and the difference is that its appearance can be assembled at will with the help of 3D printing technology.

Figure 3-26 3D printed UAV

Today, except for the standard technical modules, such as the core printed circuit board for autopilot, motor, batteries, and camera, among other remaining components, including the propeller and its shell, are all the results of 3D printing. Users can 3D print the shape of the shell according to their preferences to form their aircraft.

With the aid of 4D printing technology, the aircraft will bring fundamental changes that could further expand the scope of UAV applications. For example, in the event of a change in the patrol environment, a UAV can self-assemble to change its shape in response to those changes (Figure 3-27). While being in an open environment, a UAV can appear and exist in a broader form. But, when entering some narrow spaces, it can self-assemble to change the shape into either circular or superimposed, allowing exploration of some deep wells or unknown caves.

In the face of different changes in weather conditions, 4D printed aircraft could also undergo self-assembly and self-change, by always keeping the camera in the most rational working state to adapt to flying and performing aerial photography tasks. Of course, these functions can also be achieved by giving the aircraft electromechanical component technology. But, with 4D printing technology, the aircraft could self-drive and self-change through its material structure rather than being driven by its motor components.

Figure 3-27 Aircraft with different shapes

Besides, the aircraft will be further impeded from moving forward in the air during its operation due to the separation of aerodynamic forces in the direction of the airflow, which is what we usually call resistance. The different parts of the aircraft will be subject to pressure drag, frictional resistance, and induced drag, caused by various factors. The pressure drag can be described as the dust that often rises behind the car while driving at high speed because the air pressure in the vortex area behind the vehicle is small, and it sucks up the dust. Frictional resistance is the air flowing against the surface of the aircraft, and due to the viscous nature of the air, it results in friction with the aircraft's surface and creates frictional resistance that impedes the aircraft from moving forward. The generation and magnitude of this resistance are related to the shape of the aircraft.

Through 4D printing technology, objects made of smart SMP will have the ability to self-fold, which will allow the unmanned aircraft to choose the optimal shape to resolve the various resistance obstacles efficiently under different flight conditions. This will create an optimal flight. For instance, it could have one shape at takeoff, one shape during the flight, and another most suitable shape when diving. This will not only minimize the air resistance to the aircraft but also reduce friction loss to ensure the best use of the aircraft.

SECTION 11 | 4D PRINTED ELECTRONIC PRODUCTS

Along with the development and manufacturing technology of electronic products that are continually improving, represented by mobile phones, computers, combined speakers, and so on, all have taken the road to miniaturization. In early 2007, with the birth of the first iPhone, electronic products that once were segregated in their functions were now heading toward gradual intensification. More and more devices started to integrate a variety of applications, and some even became all-around kings.

With the advancement of 3D and 4D printing technologies, what kind of changes are coming to electronics in the future?

First, new manufacturing grafted with 4D printing technology will make electronic products invisible. The new materials used in 4D printing will have flexible elements and memory functions, which will enable manufacturers to embed the programming of the functional features into the memory materials at the early stages of design and then incorporate the triggering media for different functional features via 4D printing. This will then allow the products to present different functions of applications by triggering their respective medium.

An electronic device that needs to be idle for a while could be attached anywhere, like our hair, glasses, watches, clothing, or shoes—a place that is almost undetectable by others. Once one of the functional applications is required, it can be activated by triggering the corresponding medium, which can be water, electricity, or external physical action, or it may even have brain waves so that its functions start working as expected. In the process, the device could be further transformed if needed.

Just like our current widely used smart glasses, the 4D printed lens, which can attach to our retina, with the blink of an eye, can start the corresponding functions, such as taking photos, video, and so on. At the same time, it can also start the thought search controlled by the brain waves, accessing the Internet to stimulate the role of an e-tourist while surfing online. It could also function as a virtual reality (VR) port, sending people to the ancient times that they were previously unable to experience, or the unknown future a hundred years ahead, and maybe even the distant interstellar space.

4

LIFE

In a lab of Harvard University, a lifeless "flower," supported by the scientists' 4D printing devices and technology, magically realized the perfect self-transformation and blossomed. 4D printing technology not only gives plants specific characteristics of life forms but also allows new possibilities in the human body's functions, including innovative therapies, organ replication, reversal of aging, as well as the improvement and optimization of various life forms. Furthermore, with the advancement of the development and application of AI, 4D printing will become a new manufacturing model for AI devices.

SECTION 1 | 4D PRINTING CREATES "LIFE"

In the dimension of time, giving the ability of self-transformation to the printed objects is the primary purpose and direction of 4D printing's initial stages. Even though 3D printing is not popular yet, it does not affect the pace of scientists promoting the development of 4D printing.

At one time, in a lab at Harvard University, a lifeless flower that was able to transform and bloom on its own was created by the scientists' 4D printer.

This research focuses on advancing the development and application of materials. In the process of 4D printing, people use a variety of materials, including hard and soft materials in a fixed form. Different materials combined in different specific ways achieve bending or transformation of each part of the structure under certain conditions. So, can we produce a similar effect with just one single self-bending material?

To achieve this, Harvard scientists use cellulose fibrils and wood pulp acrylamide hydrogels to develop a special gel. When the gel is extruded from the nozzle of a 4D printer, the fibrils are aligned with the extrusion axis, creating the hydrogel composite with directional properties. In other words, the composite can extend more easily in one direction. With the application of some preset mathematical calculations and encoding, after the composite is 4D printed, it can be made to stretch, bend, twist, or curl under the action of the stimuli.

Not only that, but scientists on the path of 4D printing materials research have also found their intellectual inspiration in nature, that is, by referring to the opening of a flower, which is a complex structure of morphological transformation in nature. Scientists, therefore, mimicked the structure of plant cells. The aligned cellulose fibrils inside the plant cells limit their range of movement, so they mixed the cellulose fibrils extracted from wood pulp with acrylamide hydrogels (a gel that expands when exposed to water). When the composite was ejected from the 3D printer's nozzle, the fibrils were aligned in rows inside the acrylamide hydrogel, which means the printed object could only expand vertically but not horizontally. At the same time, based on composite guarantees, scientists have developed a mathematical model that can obtain curvilinear shapes from the crisscrossed designs and use the alignment of the fibrils to ensure that the curvature of the composite becomes consistent with the desired design.

The result of the experiment is surprising. The two flowers were printed out in the same shape. When they were submerged in an aqueous medium, five petals curled in different directions. Besides, scientists also mimicked the design of a certain orchid. After the flattened-shape petals bent, the mimicked flower was very similar to a real orchid. After fluorescent dyes were added to the hydrogels of the composite, the flower was presented in a more beautiful form for people to appreciate.

In this way, the 4D printed flowers had the characteristics of life forms which made them very similar to the real flowers. With time, different growth patterns and characteristics will be further shown based on the action of different mediums, such as a flower as a bud, in full bloom, and overflowing with fragrance. Moreover, the mediums for the printed objects can be water, air, or sunlight, or anything else that could activate the 4D printing of flowers can be regarded as the medium. Flowers can even be freeze-frame frozen at their most beautiful moment or moved on to the next life cycle as needed by the people.

SECTION 2 | 4D PRINTING'S INNOVATIVE APPLICATIONS IN MEDICINE

In the field of biomedicine, 4D printing, as an innovative treatment method, is opening up new possibilities for disease treatment, especially in the customization of medical devices and implants.

We all know that traditional medical devices and implants are usually static, but they may not be able to adapt to changes in the patient's internal environment. The introduction of 4D printing technology offers a solution to this challenge—through 4D printing technology, we can achieve the customization of medical devices and implants.

In recent years, with the development of technology, 3D printing has taken the lead in gaining application breakthroughs in the medical field.

This is mainly due to the significant demand for personalized customization in the medical industry (especially in the field of restorative medicine), and there is a rare standard of quantitative production. At the same time, small batches, high precision, as well as customizable printing based on individual differences are the advantages of 3D printing technology and are also applied in various medical fields, including the printing of limbs and organs. However, compared to 4D printing technology, it is still relatively primitive and outdated.

The best description was published in an earlier report in the *International Herald Leader*. It has the following views on the applications and changes brought by 4D printing technology in the medical field. According to the report, 4D printing technology not only plays a role under extreme conditions, but it can also make a big splash in a tiny space. For example, inside the human body, 4D printed objects become new types of human implant materials that could even replace organs.

Using its automatic shape-changing features, 4D printing technology has unlimited potential in the biomedical field. Some biomaterials companies are already involved in the development of biological heart stents with memory functions. The conventional heart stent in the medical field is usually made of memory metals, which will automatically hold open and dilate the blood vessel passages after implantation in a predetermined position through a blood vessel. But the problem is that the metal stent is not degradable, and unless it is manually removed, it will remain in the body forever, causing many complications and adverse effects.

If a new biological heart stent is developed using 4D technology, it will be able to transform as well. First, it could enter the human body in an extremely tiny form, and when the condition in the body would be most suitable, it then would expand and transform into a hollow stent that could hold open. Since it will not be made of metal, it may also be degradable. After the drug-coated stent completes the mission of dilating and unblocking the blood vessel at its predetermined position, it will automatically degrade in the blood. Similarly, the use of 4D printing technology to print out the replacement of human organs, such as joints and cartilage, will have more bionic qualities and will be able to make itself more similar to the real organs of the human body.

Clearly, every patient's physiological condition is unique, making traditional generic medical devices and implants insufficient for individualized needs. 4D printing allows doctors to design and manufacture devices tailored to a patient's specific conditions, enabling intelligent adjustments based on temperature, pressure, or other physiological changes within the patient's body. Such personalized medical devices can enhance adaptability, thereby significantly improving treatment outcomes.

Another significant advantage of this technology is that it provides patients with a more comfortable and effective treatment experience. By monitoring physiological parameters within the patient's body, medical devices and implants can adjust their shape in real-time to adapt to the changing environment. For example, implants can adjust their shape based on changes in body temperature, ensuring optimal adaptability and comfort throughout the treatment process.

Take, for instance, the condition of urethral stricture, which affects one in every thousand children, with some even facing this issue while still in utero. If urine cannot be successfully expelled, urethral stricture can burden the kidneys of the child, potentially leading to fatal outcomes. Long ago, cardiovascular stents were used to treat blocked blood vessels. A balloon wrapped around the stent is placed at the narrowed part of the vessel, and as the balloon expands, the stent opens up. Finally, the balloon is deflated and removed, leaving the stent to support the vessel. While effective, fetal urethras are much narrower than cardiovascular strictures, and previous technologies struggled to create stents suitable for fetal urethras.

In 2019, ETH Zurich indirectly utilized 4D printing technology to produce mini stents that are 40 times smaller than existing ones. They first used lasers to create a stent mold, filled the mold with memory material, and then melted the mold away. This memory material can "remember" its original shape, returning to its initial structure even after deformation. Therefore, after being compressed and inserted into the fetus, influenced by body temperature, this micro stent can autonomously expand to its original shape, aiding in the dilation of the fetal urethra. Although these microstents are not yet available for clinical testing, it is foreseeable that in the future, technology utilizing 4D printing will pave the way for more minimally invasive surgeries.

Moreover, in surgical applications, 4D printing technology offers doctors more precise and safer solutions. Customized medical devices can better adapt to a patient's anatomical structure, reducing surgical risks. By designing devices that align with a patient's physiological state, surgeons can increase the accuracy of operations, minimize disruption to surrounding tissues, and thereby speed up the recovery process. This means fewer complications and faster rehabilitation.

SECTION 3 | 4D BIOPRINTER

On December 23, 1954, the US successfully conducted the world's first human organ transplant. Since then, organ transplant technology has rapidly developed, and since the 1970s, organ transplantation has become a viable option for patients with renal failure and other organ diseases. However, due to the limited number of donated organs, many patients are forced to wait in long lines for life-saving organs. According to the World Health Organization, around two million people worldwide require organ transplants each year, but a severe shortage of organ donors means the global average supply-to-demand ratio is less than 1:20. In the US alone, over 100,000 people are waiting for a suitable organ, with more than 90,000 of those waiting for a kidney transplant. The average waiting time for a kidney is three to five years, meaning that, on average, more than 20 people die each day while waiting for an organ.

Anthony Atala, a pioneer in US surgery and regenerative medicine, once gave a speech, which was about the rising health crisis because of the increasing prevalence of human lifespan extension and organ failure. In the US alone, as many as 100,000 people are waiting for an organ transplant, and that number doubles every decade. Yet, against this backdrop, the number of organs available for transplantation has not increased and is less than one-third of the total number of people who need organ transplants. On this occasion, on the stage, Atala demonstrated a special 3D printer that could create a prototype of a human kidney. When Atala was wearing medical gloves and holding a complete kidney that had just been "printed out," the audience was deeply shocked!

If the kidney printed by Atala was still in the demonstration stages, the cardiovascular surgeon at Miami Children's Hospital, using 3D printing technology, was able to make a girl's heart with real-world application. A 4-year-old girl named Adanelie Gonzalez suffered from a congenital disease called total anomalous pulmonary venous connection, which meant her veins' were unable to deliver blood to the correct part of the heart, resulting in breathing difficulties, lethargy, and frequent illness (weak immune system). After several restorative surgeries, doctors realized that without a permanent repair, Gonzalez's life would be only a few weeks long. At that point, doctors successfully performed the surgery using a 3D printed donor heart. Now, Gonzalez's blood is flowing normally, and she is currently recovering at the hospital.

Meanwhile, 3D printing has gradually replaced infected jaws, cancerous vertebrae, and deformed hip bones in the world's leading medical community through the human anatomy. In June 2014, Len Chandler, who is now a 71-year-old from Rutherglen, Australia, was diagnosed with cartilage cancer in his heel. His surgeon contacted the experts from Australia's Commonwealth of Scientific and the Commonwealth Scientific and Industrial Research Organization, and they brought expertise from an Australian company, Anatomics, which produces and customizes medical equipment. By using the heel-bone anatomy chart provided by the company, the research team single-handedly developed an ideal implant and provided it to the surgeon from St Vincent's Hospital. The important thing was that surgical sutures could be inserted in the holes of the implant, which connected the smooth and rough surfaces of the bone to allow the tissues to grow closer together. In July, a titanium heel-bone implant (Figure 4-1) was successfully implanted in Chandler's foot. Three months after the surgery, Chandler returned to the hospital for a follow-up visit, and the doctor said that he was recovering well, and his heel was able to bear some weight.

Internationally, "bioprinters" have been used in many applications. In China, similar success stories are common. Doctors at the Peking University Third Hospital successfully implanted a customized 3D printed vertebra (Figure 4-2) to a 12-year-old boy to replace the segment of the boy's cancerous vertebra in his neck.

Figure 4-1 Titanium heel-bone implant

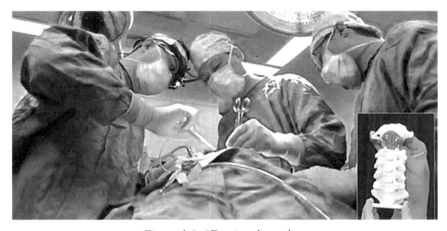

Figure 4-2 3D printed vertebra

During the 5-hour surgery, surgeons removed the cancerous vertebrae and implanted a 3D printed vertebrae between the boy's first and third vertebrae. The surgery involved removing nerves, carotid arteries, and cancerous tissue from the spinal cord and fixing the artificial vertebrae with titanium screws. According to the director of the hospital's orthopedics department, "If using existing techniques, the patient's head needs to be framed and pinned after the surgery, which means the patient cannot touch the bed while resting, and this lasts at least another three months. But with 3D printing, we can mimic the shape of the patient's vertebrae, which is a lot more powerful and convenient than traditional methods." The biggest advantage of 3D printing is the ability to customize medical implants to create artificial bones that perfectly match a patient's skeleton. Currently, the boy is in good health, and his recovery is progressing as expected.

The "bioprinter" can cultivate the patient's cells and then use 3D printing technology for replaceable skin or organs, realizing its great potential in some areas of medicine. Moreover, the development of 3D printing will eliminate the need for organ donation waiting lists. Likewise, skin transplantation will become history. Along with the application of 3D printed human tissues in drug testing, it will also reduce the need for animal experiments.

With the development and application of 4D printing technology, bio-printing will be further perfected. In the field of artificial organs, 4D printing technology represents not just an innovation in shape but also significant progress in biomimetic performance. Take the example of the movable artificial heart valve; this technology, by combining biocompatible materials, successfully simulates the dynamic characteristics of natural organs. The movable artificial heart valve can elastically deform according to the patient's heart activity, greatly enhancing its realism and physiological similarity. The successful application of this technology not only offers patients more intelligent, adaptable artificial organs but also brings new hope to the treatment of cardiovascular diseases and other areas.

In tissue engineering, 4D printing provides an effective means to control material deformation and tissue arrangement precisely. By regulating the properties and structure of materials at both the micro and macro levels, researchers are able to construct more complex, hierarchically structured tissues. This level of precision offers unprecedented possibilities for repairing and reconstructing damaged tissues. For example, in the skeletal system, the application of 4D printing technology can achieve artificial bone tissues that are closer to the natural structure of bones, providing more ideal implants for surgeries such as fracture repair and joint replacement.

People may cultivate the printed cell layer directly on the wound in the future, and it will transform into a shape as the wound evolves. An instrument could even be penetrated deep into the patient's body to detect the damaged cells and replace them with the new printed cells through a minimally invasive operation. Instruments could use their probes to repair wounds directly.

Taking bioprinting from imagination into reality allows the human body to renew cells in a timely manner and promptly repair and renew the diseased and aging human tissues. When people frequently use "face printers" to pursue eternal youth, a guy who sees a sexy girl may say, "Hey, girl, you look good!" and the girl may turn around and tell him, "Thank you, young man, I've been printed for 87 years." The young man would then look back with a light smile, "I've been printed for over a hundred years."

SECTION 4 | 4D PRINTING AS A NEW ANTI-CANCER TECHNOLOGY

According to the 2020 global cancer burden data released by the International Agency for Research on Cancer of the World Health Organization, there were 19.29 million new cancer cases worldwide in 2020, with China accounting for 4.57 million new cases, or 23.7% of the global total, making it the country with the highest number of new cancer cases. China has indeed become a "cancer giant."

The top ten countries by number of new cancer cases in 2020 were China with 4.57 million, the US with 2.28 million, India with 1.32 million, Japan with 1.03 million, Germany with 630,000, Brazil with 590,000, Russia with 590,000, France with 470,000, the UK with 460,000, and Italy with 420,000. There were 9.96 million cancer deaths worldwide in 2020, with China's cancer deaths reaching three million, accounting for 30% of total cancer deaths, mainly due to the large number of cancer cases in China, placing it first in the world for cancer mortality.

The battle against cancer is a long and enduring struggle for humanity.

Cancer has existed since ancient times. The "Tianguan" of the *Zhou Li* from the Western Zhou Dynasty records the treatment of tumors, ulcers, surgical wounds, fractures, and other diseases, indicating that cancer is not a product of modern civilization but one of humanity's oldest enemies.

For thousands of years, humanity has sought a cure for cancer, yet we still lack effective medical technology against it. Even the most common chemotherapy, while effective in destroying cancer cells, also harms immune cells. However, 4D printing technology may change this situation.

Through some of the previous chapters in this book, we received some understanding of 4D printing technology. The entry of 4D printing technology into the medical field, or the field of printing cells, will bring a fundamental revolution to the current medical technology. That is to say, we can mimic white blood cells and 4D print them with biomaterials. Of course, the key to this lies in printing materials, which will have great uses in the medical field.

We can simply understand that a 4D printed cell "robot" is an anti-viral nanorobot made from materials that are designed according to DNA strands in the form of a clamshell or the same as a normal human cell. When such robot cells are input into the blood and come into contact with specific cancerous cells, they will release specifically targeted antibodies that halt the cells' growth or mimic the behavior of a specific form that attacks the specifically targeted cells.

The medium that triggers the cell robots is the cancerous cells. The triggering medium could be any specific kind of cancerous cells or a variety of cancerous cells. While the 4D printed robot cells are flowing in the blood and patrolling the human body's white blood cells, once they detect the cancerous cells, the robot cells will be able to target and attack and exceeds the killing power of the white blood cells (Figure 4-3).

For patients who have had cancer, we can inject the 4D printed targeted robot cells into the cancer site to attack or devour the cancerous cells. This will effectively improve the current treatment methods for cancer and provide multiple new solutions for humans to overcome cancer.

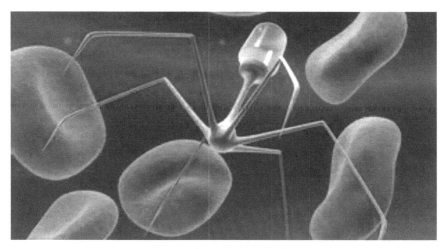

Figure 4-3 Attacking cells

- SOLUTION 1: 4D PRINTED "GUARDS" FOR THE HUMAN BODY

With the support of nanotechnology, 4D printed non-therapeutic nanorobots will be able to function as the guardians of the human body and conduct 24-hour patrolling inside the human body.

On the one hand, the "guards" can timely clean up the residual "garbage" in the body, especially in the blood vessels, and the "garbage" could then be excreted with metabolism. On the other hand, what is more important is that the "guards" can also detect cells with cancerous potential in time and issue a warning immediately or directly kill the cells before they turn into cancerous cells to ensure the stability and harmony of the human body.

- SOLUTION 2: 4D PRINTED ANTI-CANCER DRUGS

4D printed objects with the ability to self-transform have unlimited potential in tiny spaces. In the process of cancer treatment, we can replace or repair damaged cells and tissues with 4D printed organs or stents.

Taking the biological heart stent as an example, the new 4D printed biological heart stent could enter the body in an extremely tiny form, and then expand and transform into a hollow stent that could be held open under a suitable environment in the body. The stent is like the 4D printed underground water pipes that can adjust the capacity and flow by expanding or contracting the radius of the pipes. It can even self-repair when damaged or decompose when "scrapped."

• SOLUTION 3: 4D PRINTED HUMAN SKIN

Scientists in the Netherlands have recently actualized the use of stem cells as ink for 3D biological printing of human skin, which is regarded as the prelude to 4D printing of human skin. Relying on the self-transformation characteristics of the editable materials, 4D printed skin that fully integrates the features of individual skin textures will achieve the greatest degree of fit in the process of replacing cancerous or burned human skin. This can dramatically improve the healing rate of carcinogenesis or other dermatological patients.

• SOLUTION 4: 4D PRINTED LIVING TISSUE IMPLANTS

Earlier, researchers in Melbourne have found a way to generate a kind of cartilage that can "self-generate" for cancer treatment and replace damaged cartilage. On this basis, scientists can use editable materials and incorporate them with the genetic parameters of the human DNA strands while editing. Upon that, the living tissue implant obtained by 4D printing the materials will minimize the rejection by the human body. In case of adverse reactions, the implant can also self-transform in line with the actual environmental conditions in the human body to come out with the best solution to adapt to it.

• SOLUTION 5: 4D PRINTED MEDICAL EQUIPMENT

The process of cancer treatment is a life-saving process but also a life-threatening one, at the same time. This is because modern medical treatment kills many healthy cells that are useful to the human body in the process of killing cancerous cells through radiotherapy. In this treatment process, if cancerous cells can be isolated and the radiation area can be more precisely located, the cancerous cells can be killed effectively without damaging the human body. This will help in improving the success rate of cancer treatment.

Medical equipment like 4D printed radiotherapy auxiliary will play an active role in this aspect. These 4D printed medical equipment machines can enter the human body in small volumes and transform according to the living environment of different parts of the human body to isolate cancerous cells effectively and protect healthy areas so that the cancer treatment becomes "harmless." This is especially important for tumor treatment of some vital organs or vulnerable areas, such as the nose, eyes, ears, etc.

Implanting 4D printed objects into the field of biomedicine, particularly expanding these to human body applications, is the gospel for the development of human health and medical treatment. However, even though we are excited by the prospect, we must also be calmly aware of the risks involved, and we must not underestimate them. Let's say the 4D printed cells or nanorobots wandering around in the human body can easily turn into prototypes of biological weapons used by people with bad intent if the monitoring is not strict. The self-transformation feature of the editable materials makes it even riskier for criminals to use compared to 3D printing. For example, forbidden objects like guns. The 3D printed guns are concrete and direct, which is relatively easy to detect and control, while guns that are printed via 4D printing can take any form, and only under certain environments and actions of the triggering medium will they transform into their predetermined shape. This is unpreventable.

SECTION 5 | 4D PRINTING IN REVERSING AGING

Aging is an eternal topic that humanity cannot avoid. Since ancient times, people have been trying to alter the process of aging. Previously, Emperor Qin Shi Huang engaged in massive construction projects and believed in the myth of eternal youth, even sending Xu Fu along with five hundred boys and girls overseas in search of an elixir of life. Later, Emperor Wu of Han also sent envoys to seek immortality and built high platforms to collect so-called immortal dew.

Today, with the development of medical science and technology, the average human lifespan has significantly extended over the past few centuries. In 1900, the global average life expectancy was only 31 years, and even in the wealthiest countries, it was less than 50 years. By 2015, the average human life expectancy had increased to 72 years, and in Japan, it even reached 84 years.

This undoubtedly makes the human desire for youth, health, and longevity even stronger. However, we must acknowledge that despite centuries and even millennia of exploration, humanity has yet to invent an elixir of immortality.

Yet, even facing such a difficult problem, 4D printing can offer new solutions. In fact, looking at the evolution of lifespan, the progress of life science technology has been the most important driving force behind the increase in average lifespan from the "high age" of 40 years to today's average of 80 years.

First, the most direct factor affecting human life expectancy is the advancement of modern medical diagnostics and treatment technologies, each of which contributes to longer, healthier lives. So, what can 4D printing do in this link?

At its simplest, the field of pharmaceutical science can use 4D printing to manufacture drugs that provide the most timely and effective disease control and treatment. For example, through the appropriate design of some polymers, a drug delivery vehicle can be created to deliver the drug into the human body.

The trigger medium to activate the drug in the human body is body temperature. That is, once the body's temperature rises to the critical point of fever, the drug contained within will be released and promptly control the disease. Concurrently, the drug vehicle without the drug will be excreted with the body's metabolism.

By the same principle, the medical science community can also design targeted inducing mediums according to the characteristics of their conditions so that drugs can be delivered to the corresponding locations in the human body, and once the pathological symptoms appear, the drug can be triggered and released. Thus, diseases could be perceived before they show symptoms, and the most timely medical treatment and rescue plan could be achieved.

When the prevention and treatment of diseases are realized by 4D printed drugs, the future of human science will be further committed to 4D printing in R&D and the application of technology for human organs. It will thereby enter a new era of "organ replacement" and opening up new life channels. According to the latest news, a team of experts from Xijing Hospital of the Fourth Military Medical University, in conjunction with a resident national key lab, adopted 4D printing technology to successfully treat and save a 5-month-old infant who had congenital heart disease with bilateral trachea severe stenosis that was life-threatening due to potential further tracheal damage. The 4D printed external tracheal stent tailored to the shape of the infant's trachea was made of degradable material, which not only met the performance requirements of shape, stiffness, strength, and elasticity of an external tracheal stent but also conformed to special requirements for human biocompatibility and biodegradability. The stent could be gradually degraded and absorbed by the human body in the next two years, saving the infant the pain of a second surgery to remove it. The infant's tracheal tube was removed on the same day after the surgery, and on the fifth day, cardiac ultrasound and chest CT scans showed that the cardiovascular malformation was completely corrected, and the bilateral trachea recovered well. This is the first time in the world that a 4D printed external tracheal stent was successfully used in the treatment of an infant with complicated congenital heart disease with severe bilateral tracheal stenosis.

In the future development of science and technology, through 4D printing, what humans can replace will no longer be limited to the diseased organs, but also the weakened, aged organs and auxiliary devices for human functions. Perhaps, there could be a new drug to keep humans young forever and thus compose the legend of humankind's immortality. Besides, the scientific community has already made initial attempts in the lab to grow cells in a flat framework and then for them to subsequently change the shape of the tissues.

SECTION 6 | 4D PRINTING DEVELOPMENTS IN BIOMEDICAL FIELD

4D printing is an integrative technology, based on shape-changing materials and 3D printing technology. SMPs, widely used in deformable materials, have significant applications and potential in the biomedical field. SMPs are characterized by being lightweight, highly resilient, having mild recovery conditions, biodegradable, and low or non-toxic. These outstanding properties have led to the widespread use of SMPs in the biomedical field, such as in sutures, dental braces, and aneurysm occlusion devices. However, these SMP structures are mostly simple linear structures. Compared to complex, personalized, and precision-required structures like heart stents, bone scaffolds, and tracheal stents, traditional fabrication techniques fall short. The emergence of 4D printing technology has filled this manufacturing gap.

Bronchial stents. In treating bronchial tracheomalacia, Chinese researchers used SLA 3D printing technology with polycaprolactone (PCL) as the printing material to fabricate a tracheal stent successfully. They created a 3D tracheal model based on the patient's CT scan images and medical digital imaging technology and designed a 3D graphic in STL format. By combining computer simulation with the trachea and stent model and surgically implanting it into the patient, they successfully cured three patients. Post-operative internal tests showed that the tracheal stent could biodegrade as the patient grew. Once the patient's trachea was fully developed, the material could also biodegrade, thus avoiding the need for multiple surgeries. Based on this achievement, using SMP materials with shape memory effects for tracheal stents through 4D printing will see widespread application.

Cell scaffolds. Cell scaffolds formed through 4D printing can promote cell growth and differentiation. Chinese researchers used a novel renewable material—acrylated epoxidized soybean oil (AESO), which, compared to traditional PEGDA, can enhance cell adhesion and proliferation. By changing the orientation of fibers in the scaffold, they prepared two cell scaffolds with different fiber orientations. The results showed that the deformation of the cell scaffold could mechanically stimulate the cells, leading to the oriented growth of cells and cell nuclei. Starting from the internal structure of the scaffold, researchers prepared scaffolds with biomimetic gradient pore structures, which could encourage cells to grow toward the pores, and these pores could also act as channels for nutrient and waste metabolism. These three different research examples, from various perspectives, verify the excellent performance of 4D printed AESO cell scaffolds. Currently, the types of SMPs suitable for 4D printing and possessing high biocompatibility are still limited. Further research into 4D printing technology and highly biocompatible smart biomaterials can promote the design and development of new functional biomedical scaffolds.

Vascular stents. Researchers from the Harbin Institute of Technology in China created a shape memory composite material by adding magnetic Fe_3O_4

nanoparticles to PLA. Experimental results showed that spiral stent structures made from this material could automatically expand under the influence of a magnetic field, completing the entire expansion process in just ten seconds. This self-expanding vascular stent can be used to treat cardiovascular diseases caused by thrombosis. When the self-expanding vascular stent reaches a narrowed blood vessel, its diameter can be increased and supported by adjusting the intensity of the external magnetic field, thus restoring normal blood flow. This technology not only achieves intelligent remote operation of medical devices but also offers new possibilities for minimally invasive surgery. It holds great application prospects for the implantation of smart and customized devices in the human body, significantly impacting the further development of the biomedical field. Clearly, this type of vascular stent, which can automatically adjust according to the blood vessel environment and blood flow rate, showcases the advantages of 4D printing technology and the true value of vascular stents.

Bone scaffolds. Chinese researchers used FDM technology to mix PLA and hydroxyapatite (HA) in a 20:3 mass ratio to manufacture a porous scaffold for bone defects, which possesses shape memory functions. Experimental results revealed that the porous PLA/HA scaffold could adhere well to mesenchymal stem cells (MSCs), supporting their survival and stimulating their proliferation, making it crucial for medical applications. Additionally, the presence of MSCs in the scaffold aids the formation of blood vessels at the implantation site. This shape memory scaffold not only promotes the growth and proliferation of MSCs but also has significant potential for applications in bone replacement and adaptive implants. This means that bone scaffolds printed using 4D printing technology not only encourage the growth and proliferation of MSCs but, more importantly, the self-changing materials formed by 4D printing technology can adapt according to the patient's growth, ensuring more effective growth and proliferation of MSCs.

Cardiac stents. Cardiac stents are a common surgical treatment for heart diseases, and researchers from China, combining FDM and medical technology, successfully developed a stent for intracardiac valve reshaping surgery. This stent, based on 4D printing technology, features a mesh structure design, allowing it to contract to a certain extent and expand after implantation to regain its original shape, making it particularly suitable for pediatric patients. Mechanical performance tests indicated that these stents, printed using 4D printing technology, have mechanical properties comparable to those of commonly used nickel-titanium alloy stents and also possess biodegradable characteristics. These shape memory cardiac stents prepared using 4D printing technology have broad application prospects in the biomedical field.

Dental orthodontics. Increasingly, young people and students opting for aesthetic improvements are choosing invisible orthodontic aligners (clear braces). Typically, dozens of these aligners are custom-made for each individual, requiring replacement every 1–2 weeks to gradually adjust the teeth's alignment. However, this method, necessitating multiple aligners, can be inconvenient and wasteful. Is there a better technological solution? Can a single aligner achieve the correction? Scientists from Germany, Egypt, and the UAE, leveraging 4D printing technology, have developed a novel type of transparent dental aligner made from a polymer with "4D" shape memory capabilities. This aligner, made using Kline Europe's transparent ClearX resin through the DLP process, softens upon insertion into the mouth, adapting to and repositioning the teeth, then changes as needed for correction. Traditional methods move teeth only 0.2 to 0.3 millimeters per aligner change, or rotate 1° to 3°, requiring about 14 days per aligner, thus extending the correction period and increasing costs. The new material and the biological considerations of tooth movement address the main drawbacks of traditional methods. SMP, recently introduced into dentistry, especially for orthodontic applications, offers vast potential. The inherent advantage of 4D printing technology in applications requiring shape change over time presents a natural solution for orthodontics, promising future applications in this field.

Novel capsules. MIT researchers have developed a micro-scale drug capsule using 4D printing technology that changes shape with temperature. When the body experiences symptoms of fever due to illness, the capsule deforms, releasing its medication. This temperature-driven mechanism allows for more controlled timing of drug release, providing immediate medication before the body recognizes the temperature rise. Further advancements in 4D printing materials could lead to drugs that are released upon encountering cancer cells within the body, revolutionizing cancer treatment.

Cosmetic surgery. Researchers from the Fourth Military Medical University's Xijing Hospital in China have developed a biodegradable material using 4D printing technology. This material was used to print a biodegradable breast implant for breast reconstruction surgeries. Long-term follow-up showed excellent compatibility with patient tissue, with autologous fibrovascular tissues gradually integrating and replacing the implant over two years. This 4D printed biodegradable implant design not only avoids issues of material residue in the body but also maintains breast shape, enhancing patient quality of life.

The various research and applications indicate that 4D printing technology is breaking traditional medical device and treatment limitations, offering new possibilities for minimally invasive surgery, reducing the number of surgeries, controlled drug release, precise drug delivery, and tissue or organ replacement. Moreover, 4D printing technology can quickly and accurately provide personalized medical services based on individual patient conditions, offering custom treatment plans to alleviate their suffering and improve life quality. This technology offers a new direction for the further development of biomedical science. As more SMMs suitable for bioprinting are developed and 4D printers continue to evolve, more personalized smart medical devices will be applied in the biomedical field. Therefore, the integration of 4D printing technology with the biomedical field is set to become a new trend in medical development.

5

MILITARY

4D printing not only demonstrates immense potential in manufacturing and life sciences but also holds significant value in military applications. Intelligent equipment and components can provide the armed forces with enhanced adaptability and flexibility, thus improving combat effectiveness. Additionally, the construction of military bases and facilities can achieve more flexible and variable structures through 4D printing technology, enhancing concealment and security. The application of 4D printing is expected to introduce novel solutions to the military field, propelling the development of modern warfare.

SECTION 1 | 4D PRINTING FOR MILITARY WAR

4D printing is currently focused on and applied not only in the field of people's livelihood but also in the military industry, which has shown greater interest in it. According to the US Army's Chief Technology Officer, Grace Bochenek, 4D printing "takes 3D printing and adds a transformation dimension." The idea behind this is that the properties of a 3D printed component can be altered when exposed to different environments, such as water or extreme temperatures. The applications of the idea include body armor and equipment for soldiers.

The application of 4D printing technology in the military field is receiving significant attention. According to a report by Allied Market Research in August 2022, the global military 4D printing industry is projected to reach a value of US$673.4 million by 2040. Additionally, the American market observation website reports that, from 2022 to 2031, the compound annual growth rate of the military 3D and 4D printing market is expected to exceed 10%. This underscores the importance of 4D printing technology in defense strategies.

Adaptive camouflage combat uniforms are among the early military applications of additive manufacturing technology. These uniforms can automatically adjust their color according to environmental changes, similar to a chameleon, thereby enhancing camouflage effects. The advent of 4D printing technology and the development of smart materials further enhance the camouflage function of such combat uniforms, potentially revolutionizing covert operations on future battlefields.

Weapon and equipment components designed and manufactured using 4D printing technology can undergo structural and functional changes in response to rapidly changing battlefield conditions. This enhances the environmental adaptability of the equipment, optimizes performance, and reduces costs. NASA has proposed a design concept for a future intelligent morphing aircraft whose shape can adaptively change according to external conditions, such as modifying its span length to optimize lift-to-drag ratio for extended range and flight time or altering wing curvature to enhance maneuverability, thus improving combat performance. Canadian researchers have also used 4D printing technology to develop a new type of adaptive, flexible wing for drones, which can improve the wing's flight efficiency and reduce manufacturing costs.

The self-assembly capability of 4D printing also has a wide range of applications in the military field. For example, equipment such as camping tents, individual lifeboats, and battlefield medical supports can be stored in a compressed or folded state after printing and automatically expand into a pre-designed shape for use. This significantly simplifies the assembly process, reduces the cost of assembly parts, and makes them easier to carry and transport.

As a rapidly emerging additive manufacturing technology, 4D printing has achieved innovative applications combining new materials, new processes, and new mechanisms, driving the integrated dynamic design and manufacturing of "material–structure–function," and promoting the intelligent transformation of manufacturing methods. Although 4D printing still faces challenges in the types and performance of smart materials, printing processes and equipment, and the assessment and inspection of intelligent components, the future of 4D printing technology remains promising.

Bochenek said body armor could be 4D printed in the future. The Army has been struggling over the past decade to meet the dual needs of body armor, which is to be both protective and light enough not to burden the soldier or restrict their movement. In the future, scientists may develop lightweight and compact body armor using 4D printing that is easy to store and carry, but that can also expand and provide full protection.

Moreover, the application of 4D printing in the military field will affect serious national security missions. An article in *Science and Technology Daily* on "4D Printing Technology's Broad Prospects for Military Applications" will focus on the impacts of military technology and the issues of national security brought about by 4D printing. The article's points can be summarized as follows.

While 3D printing technology is on the rise, 4D printing has come into our reality. In the movie *Transformers*, the scene where a car instantly transforms into a giant robot is just amazing. With the help of 4D printing technology, this science fiction may probably be realized in the near future. At the TED conference held in Los Angeles in February 2013, MIT researcher Skylar Tibbits placed a composite strand containing water-absorbent smart material in the water, and the strand automatically twisted and transformed into the shape of the word "MIT." As already mentioned, this was the first public demonstration of 4D printing technology, and it became a sensation around the world.

4D printing technology is about the refinement of 3D printing technology. 3D printing is an additive manufacturing technology that is based on digital model files, and it uses powdery metal or adhesive materials to construct an object by printing it layer by layer. Compared to 3D printing, 4D printing adds a dimension that can "change" the object, and its core technology is the self-assembly of materials. This means it can embed certain smart materials in predetermined positions during the 3D printing process, then place the printed object in a specific environment, and physical or chemical changes will then occur under the influence of the external environment. These changes will lead to further changes in the elements of the object, such as altering its overall shape, stiffness, and so on.

Today, the broad application of 4D printing technology has attracted a lot of research in the economic fields. It also has huge potential prospects in the military fields.

To open up the link between the manufacture of weapon equipment. 4D printing technology will overturn the traditional life cycle of weapon equipment and possibly optimize it from manufacture to deployment to use, scrap to manufacture, and back to deployment, site modeling, use, recycling, and then redeployment. At the deployment site, according to different surrounding environments and operational goals, weapon equipment will be able to enhance and adjust its design parameters and be rapidly molded. It could even self-adapt to the environment to significantly improve its environmental adaptability and operational effectiveness. For instance, through 4D printing technology, it is possible to obtain a camouflage net, which will automatically change shape and color in response to the changes in external lighting. It could better blend with the surrounding environment and improve the camouflage.

To facilitate the on-site manufacture of large equipment components. Although the prospect of 3D printing is broad, it is limited by the size of 3D printers. It can only actualize the on-site printing of components in small and medium sizes. With 4D printing, the folded form of a large structure, the essential parts, and the sensitive materials required for its unfolding can be designed in advance. Then, a semi-finished component could be printed using a 3D printer, and by the control of the special physical field, the finished component could be unfolded automatically. The typical application of this is the space required for the self-assembly of large structures such as satellite solar panels, antennae, etc. This will significantly reduce the number and weight of mechanical parts needed, lowering down the volume and weight required for satellite launches.

To promote the development of micro-military robots. The microrobots will perform a large number of reconnaissance or strike missions on the battlefield in the future. The volume, weight, and energy consumption are very much related to the mechanical components required for robot movement and transformation, such as gears, chains, and so on. The 4D printing technology will provide a new technical solution for microrobots' movement and transformation. That is, the precise design and control of sensitive materials will be expected to replace traditional mechanical components. Gears are a good example. They could achieve the movement of robots, thereby considerably reducing the robots' weight and energy consumption requirements.

To reform the military logistics support process. 4D printing technology allows more weapon equipment to be manufactured in a folded form, reducing the volume of the equipment, facilitating long-range maneuvering, and decreasing the number of unnecessary damages that may occur during long-distance transportation. The 4D printed semi-finished components will have stronger malleable capacity and environmental adaptability, which can also reduce the variety and inventory of the equipment, improving logistic efficiency and enhancing operational effectiveness.

4D printing has a profound impact on the military field. Particularly for some technological explorations led by the US, including the applications of body armor and ordnance, as well as facilities like camouflage techniques and outdoor communication systems—all will be affected by 4D technology.

SECTION 2 | THE CHANGING FUTURE OF WARFARE

The emergence of disruptive technologies often first finds application in the military domain. While 3D printing is burgeoning in the new military revolution, what kind of future warfare might 4D printing bring?

"Low-Cost" Warfare

4D printing will drive cost-effective military production. Whether it's manufacturing and assembling large weapons systems such as submarines, space stations, and spacecraft or designing and producing high-end small weapons systems like smart protective gear and special firearms, research, and development, as well as production costs, are high for research units and military factories. The advent of 3D printing has partially solved the initial problems of manufacturing and production. Professor Shen Dingli of Fudan University metaphorically compares the principle of 3D printing military weapons to traditional manufacturing being a sculpting art that subtracts unnecessary parts, while 3D printing adds only what is needed. 4D printing technology goes a step further by achieving variable assembly of different weapon components on the basis of integrated printing.

Thus, 4D printing simplifies the cumbersome processes in industrial manufacturing. Designers need only to program simply without the need for upstream production equipment like machine tools, making the difficulty level of parts and product structures no longer significant. BAE Systems in the UK revealed that integrated printing of certain components of the "Typhoon" fighter jet is expected to save £1.2 million in costs. The industry upgrade brings not just savings in funds but could potentially revolutionize cost structures. For instance, weapons and accompanying equipment on aircraft carriers, instruments on satellites, space stations, and spacecraft face astronomical losses if a component fails with no materials for repair. 4D printing technology fundamentally solves the problem of material placement in "every inch is precious" special environments.

"Rapid" Warfare

Rapid warfare is first reflected in the production and manufacturing of weapons and equipment. Undoubtedly, military production under 4D printing will significantly shorten the weapon upgrade cycle, breaking through the barriers to military innovation time. Traditional weapon development usually involves simulation before manufacturing or adjusting the simulation effect while constructing. The early design and production involve organizational division and coordination of personnel, transportation, and assembly of components, while later adjustments involve remaking some components, reorganizing equipment, or even core changes. Although the entire process is highly integrated, it is somewhat dispersed and lengthy, inevitably prolonging the R&D cycle of weapons and equipment. 4D printing fundamentally changes rapid modeling: the tight integration of hardware and software will revolutionize the traditional demand-oriented—design experimentation—manufacturing R&D process, integrating design modeling to product formation closely, without considering complex model structures or relying on high-end auxiliary equipment, significantly reducing the time required for product completion. This will undoubtedly accelerate the cycle of updating and replacing weapons and equipment, allowing for the transition from mass production to customized manufacturing.

The guarantee of "rapid" production and development, in turn, promotes the "rapid" nature of warfare. The quick transformation of the "needs" and "demands" characteristic of 4D printing will meet the actual combat requirements of the battlefield. The battlefield environment is complex and changeable, and equipment damaged in battle is difficult to repair promptly, affecting the normal execution of combat missions. With the widespread application of 4D printing on the battlefield, it can effectively solve the problem of rapid transportation and repair in long-distance complex environments, even achieving real-time self-repair of weapons and equipment. Even if weapons and equipment are damaged beyond repair, with 3D printing, components can be quickly "manufactured" on-site with three-dimensional simulation graphics.

"INTELLIGENT" WARFARE

The development and application of 4D printing technology rely more on the progress of smart materials rather than the printing method itself. "It requires not just any ordinary materials, but intelligent materials with memory functions, a new type of functional materials capable of sensing external stimuli and self-transforming or assembling based on judgment." The continuous expansion of combat missions in information warfare imposes more special requirements on weapons and equipment. Relying solely on the inherent performance of weapons and equipment can no longer adapt to the changing battlefield environment. This demands that new weapons should make appropriate "dynamic" adjustments based on different battlefield environments. For example, if parts of a fighter jet made using 4D printing technology are damaged by the enemy, the jet need not make an emergency landing nor wait for the delivery of parts and the arrival of technical personnel. The damaged part will quickly detach from the plane, and new components can self-reconstruct. This intelligent "regenerative" capability will strongly guarantee the completion of more intense combat missions.

Sofas can self-assemble, and pipeline routes can be intelligently laid underground. The intelligent application of 4D printing technology is upending certain traditional civilian industries. Against the backdrop of military-civil fusion, some successfully transformed cases can be directly applied to modern military logistics. Modern military logistics consists of many links, including supply and demand, finished product production, material allocation, transportation, and classification distribution, involving factors such as transportation, weather, and terrain, making the process complex and difficult to control. Precision equipment or fragile materials might accidentally be damaged during the equipping process, causing inconvenience to logistic support. Intelligent products from 4D printing technology, which are less affected by external factors and are capable of automatic assembly into finished products, can better solve these issues. In terms of transportation, weapons, and equipment can be made in a folded state or semi-finished products and intelligently self-assemble when needed. This simple change reduces the volume of transport equipment, lowers the possibility of damage, and facilitates long-distance maneuvering.

"Personalized" Warfare

The application of various intelligent weapons will inevitably lead to personalized warfare. The innovative combat models of information warfare present more special requirements for combat, relying on the inherent attributes of a combination of weapons and equipment that can hardly meet the diverse battlefield needs. This requires the combat forces, including weapons and equipment, to make "personalized" adjustments according to different battlefield "scenarios," creating "unique" combat forces. 4D printing technology not only allows rapid printing of ideas from the human mind but also presets various possible correction elements in the printing material scheme. Therefore, after the weapon is printed, based on different battlefield situations, users can drive the weapon to self-transform, refine, correct, and "customize" targeted weapons according to their ideas. Personal weapons are relatively simple; perhaps just changing a parameter setting can achieve unexpected effects. For large system-type weapons and equipment, 4D printing needs to solve the problem of self-assembly at a higher level, add self-creative change stages, and help users achieve "unity of man and weapon." During the application process of weapons and equipment, design parameters can be optimized and adjusted according to the surrounding environment and combat objectives, rapidly shaped and formed, even environmentally adaptive, thus greatly improving the environmental adaptability and combat effectiveness of weapons and equipment.

Undoubtedly, as one of the representatives of disruptive technologies, the application of 4D printing in the military field will become increasingly widespread. Its influence will continue to deepen, and it will inevitably trigger an epoch-making military technology revolution and combat mode transformation.

SECTION 3 | WALKING IN OUTER SPACE

4D printing has great importance in the military field. Also, it will bring major technological changes and breakthroughs to the aerospace industry. As humanity's footsteps in exploring outer space continue to move forward, the question of how we could travel among the galaxies more efficiently and in a controlled manner will become the direction of a new round of exploration in the scientific community.

Landing in space, exploring the future. A manned spacecraft is not only a ladder for humans to go into space but also a bond between space and humans—it carries countless possibilities and hopes. The traditional manned spacecraft we currently operate is often composed of two capsules. One is a sealed capsule for astronauts, which is equipped with a life support system that can guarantee their water and air supply, an attitude control system that controls the attitude of the spacecraft, a beacon system for measuring the spacecraft orbit, a parachute recovery system for landing, and an ejection seat system for emergency life-saving. While the other one is the equipment capsule, which contains a braking rocket system that separates the manned capsule from the spacecraft orbit and returns to the ground, a battery that supplies electrical energy, an oxygen cylinder, nozzles, and other systems. The two capsules work together to complete the astronauts' round journey to space.

At the same time, because of the great difficulties in spaceflight operations, scientists usually take multiple measures on manned spacecraft to prevent and address some particular problems.

First, they use environmental control measures. Their main function is to adjust the temperature, humidity, and pressure inside the capsule and the spacesuit to ensure the amount of oxygen, ventilation, and water needed by the astronauts, and to absorb and dispose of waste.

Second are manual control measures. These are relevant mainly in the event of a failure of the automatic system, and when humans take over the operation and control the spacecraft to handle the emergency and avoid accidents.

Third, safe return measures. This is done mainly to ensure the safe return of astronauts. Other than providing a reliable heat protection layer to keep the return capsule from burning up, it is also necessary to make the braking overload device effective enough during the return process to guarantee that the astronaut's body can withstand the landing. Meanwhile, improving the accuracy of the landing point allows the astronauts to be located in time.

The spacecraft itself and the matched precautions are closely related to each other. More equipment and a more complex setup will undoubtedly be accompanied by an increase in the risk. 4D printing will be a new solution, and we hope to maximize the simplicity and safety of the journey to space.

4D printing will enhance the astronauts' body functions. 4D printed high-energy food and drink could be implanted inside the human body, and along with the time and feedback from the human organism's needs, the food and drink will auto-release and dissolve in an orderly manner for the human body to absorb them. Besides, 4D printed microrobot doctors will be patrolling the astronauts' bodies 24 hours a day to collect physiological information and to safeguard the astronauts' health and safety. Furthermore, there will be more 4D printed human-assisted functional devices that will mimic survival conditions in outer space and carry out environmental design and code programming accordingly. This will mean that the astronauts wearing these devices can move freely in the spacecraft and even outside in outer space.

When we are exploring space in the future, astronauts and scientists will no longer worry about gravity unbalance as they do today. Under different atmospheric and gravity conditions, the 4D printed clothes will automatically adapt by changing their shape and size. Their shoes will also self-adjust for different landing contact surfaces to increase the grip or friction, ensuring that the astronauts can conveniently walk.

In the meantime, 4D printing technology will also make the manned spacecraft more able to self-transform. During the transport of astronauts to space, the function of the manned capsule will play the main role, while after the astronauts complete the space mission, the spacecraft will perform a function conversion; that is, the function of the equipment capsule will take over to send the astronauts back to the earth.

Not only that, the most significant difference between the 4D printed spacecraft and our Vostok-1 spacecraft is that the Vostok-1 and the launch vehicle are all one-time use spacecraft. They can only perform one mission. However, a 4D printed spacecraft and equipment will enable continuous reuse through their self-transformation and self-recovery mechanisms. On this basis, constant refinement and upgrading of equipment will be achieved.

4D printing technology and materials, besides being applied in spacecraft and aerospace, will also play an active role in the construction of space stations and spacewalks. With the upgrading of space equipment construction pushed forward by 4D printing technology, humankind will have the ability to step out of the earth and conveniently walk up the stairs to outer space.

With the increasing use of smart sensing technology in the materials for 4D printing, that is, the integration of smart wearable technology will bring tremendous changes and impact to aerospace and military activities. It will thereby build up our cognition and understanding of the unknown world. Tomorrow, with the development of 4D printing technology and the realization of its applications, humankind will walk toward the "universe village" from a "global village," opening a new world of an unknowable future.

6

COMMERCE

The rapid development of science and technology will inevitably bring about changes in the existing modes and business forms of human society. So, what kind of changes will the advent of 4D printing and the advancement of its materials and technology bring to our social ecosystem? What will be the variables for the structure of today's widely used economy models like B2B, O2O, and P2P? Especially for those "ready-to-go" new sharing economy models, will they be able to join the 4D printing movement and be further boosted? Variables are everywhere, and the only thing that will not change is that 4D printing is about to open up a new model of the human business ecosystem.

SECTION 1 | THE BUSINESS ECOSYSTEM REVOLUTION

I have been thinking that with 4D printing as a new production means, our lives will experience such strong disruptions that I wonder what kind of a business scenario the future will bring. The current hot Internet Plus and the shared economy model represented by UBER has brought some profound changes to our existing business scenarios, especially to the huge business entities built in the last commercial wave. Before, we were in a period that was relative to the material shortage and technology was not as advanced as today's. Lenovo, Haier, Suning, and other huge business entities that rely on their powerful off-line channel systems built commercial aircraft carriers in their respective fields before the era of Internet Plus.

But when it comes to the era of Internet Plus, some of the key elements that once led to their success have now fundamentally changed. In a fundamental sense, the Internet does not appear as a new production means but rather exists as an information tool. However, the in-depth social changes it has brought about have transformed our current business ecosystem entirely, and more changes are happening on the edges of this ecosystem. Therefore, in the era of 4D printing, what has changed is not information tools but the entire means of production. Just as the emergence of the steam engine has changed the entire production means, production tools, and production methods, 4D printing also appears in our time in the form of such new production means and production methods.

We can imagine that once the production means, production tools and production methods are changed, the transformations they bring will far exceed the impact of the Internet Plus wave we are facing today. A simple example is that before the steam engine revolution, what were the production tools we relied on to sustain ourselves? It was nothing more than the initial hand tools of the traditional agricultural society. So, the business ecosystem of the entire society was built on a relatively slow production model at that time. After the Industrial Revolution, due to the changes in production means, production tools, and production methods, the balance between supply and demand for goods was broken, and thereby we saw the emergence of a new business ecosystem, which lasted for a long time before the Internet.

Upon entering the era of the Internet, although it did not change the entire production means as the Industrial Revolution did, it has changed the information flow of production means, production tools, and production methods. Judging from the current situation, the information flow has transformed the structure of the entire business ecosystem. So I'm thinking that with the emergence of 4D printing technology, the series of changes of production means, production tools, and production methods that it brings will be similar to another new Industrial Revolution as the one triggered by the steam engine. Under the new "Industrial Revolution," our business ecosystem and business structure will unavoidably undergo fundamental changes.

It is foreseeable that in the era of 4D printing, business forms such as Alibaba, Wanda, and the current hot O2O model will be replaced and then reappear and exist with new or unknown business terms. The huge factories and the powerful assembly capabilities that we are proud of today will be replaced and exist in a very embarrassing state during the era of 4D printing.

If 3D printing changes the manufacturing ecosystem, then 4D printing will change the entire business ecosystem. A simple example is a change in IKEA. In the future, when we own a 4D printer, we can print a few layers of material at home or in a specific place. All we need to do is give it a wash on the balcony, and after a while, a desk will automatically assemble itself and appear before our eyes. We will no longer need to assemble it by ourselves. Therefore, 4D printing will bring a revolution not only in materials but also in the business ecosystem.

SECTION 2 | NEW MODEL OF BUSINESS SCENARIOS

In the era of 4D printing, most products are not going to be produced in factories, instead, they will be produced by different 4D printers equipped in a community or several communities. Regardless of the current shopping malls or e-commerce platforms, the value and significance of the current business methods will be lost. For general products, desktop printers for home use will be installed in thousands of households. Users could then print the products they will need based on their ideas. Moreover, they could also print out some of their ideas into products and use information tools to complete transmission, transactions, etc. The role played by an individual user in the 4D printing era will be diverse. They could be consumers, users, and producers.

For some special products, especially those that are relatively large-scale, the printers used will be provided by organizations or individuals, and consumers will only need to put forward their needs or ideas on the big data platform of the Internet of Things (IoT),* the platform will then analyze the consumers' preferences and then provide them with the optimal solution. Hence, in the future 4D printing era we are waiting for, what kind of words can be used to describe the business ecosystem? At least the currently existing business terms are not accurate enough. Perhaps we can only get a sense of the new business terms when we enter the new era.

Just imagine, today we keep spending money on platforms to create the socalled "new monopoly business systems." Why don't we invest in this disruptive technology that will truly disorder the future business ecosystem? This is the future. Based on powerful technologies on the way, such as smart wearables, AI, and the entire IoT era established by them, the flow of our information will be further transformed. In addition to the changes in the entire production methodologies brought about by 4D printing technology, which I cannot simply describe in words, all I can say is that today's entire business ecosystem will be overturned much the same way it was overturned in the past.

---

* The Internet of Things (IoT) refers to the billions of electronics around the world that are now connected to the Internet to send data, receive instructions, and share data, which doesn't need human-to-human or human-to-computer interaction.

SECTION 3 | COMMUNITY + CROWDFUNDING + SHARING

The future may be nearing us in ways we currently can't recognize. If we have to find a few terms from the existing business terms to help us understand it, I will choose community, crowdfunding, and sharing.

In the days of 4D printing, like-minded friends or neighbors gathering together will be called a group according to the popular term of the day, which doesn't have to be as big as today's Internet-based community. It may be just a few people who have a common demand. For example, they are all interested in innovative home products, and these people constitute the group. Then, they could put their funds together, which is called crowdfunding, to buy a 4D printer that is suitable for printing innovative home products.

After that, it's all about sharing; that is, the crowdfunding group members can use this 4D printer according to their needs, and they only need to design the product they want to print on their intelligent terminal device. This will not be limited to a PC, but it could extend to the smart wearables, and then the design could be transmitted to the 4D printer for printing. This operation will be the same as the text printer today. Besides, the group can also rent out the idle time of this printer on an Internet platform, thereby forming a bigger shared economy circle.

SECTION 4 | CONSUMER SAFETY INDEX

A car enthusiast is longing to produce a user-customized steering wheel by 3D printing it in a basement workshop. Although the simply-printed steering wheel does not undergo quality and safety monitoring, its design file can be easily bought on a popular website that sells novel car accessories, such as novel air fresheners, gearshift handles, and others.

Imagine a car hobbyist sold a printed steering wheel online, and its buyer then installed it in their car. A few weeks later, while driving at a high speed, the buyer found that when they made a sharp left turn, the user-customized steering wheel disengaged. However, it was too late, and the car crashed, killing the driver.

Yet, the steering wheel made by 4D printing technology can avoid this tragedy. Just as the water pipelines will naturally transform in response to changes in the external environment, the steering wheel will spontaneously form the most suitable and seamless connection with the car according to the car's conditions. Ultimately, it will interact and integrate with the car, eliminating the possibility of disengagement. Moreover, if the car itself is 4D printed, it will also make the corresponding transformations in the event of a car accident to ensure the safety of its passengers.

SECTION 5 | NEW LOGISTICS FORMS

Driven by the concept of globalization, the global village has made the circulation of goods among countries and cities more common and frequent over the past decades. In particular, e-commerce companies represented by Jack Ma and Taobao have injected a booster for the transportation and circulation of goods. However, the increasing scarcity of energy and resources worldwide, such as petroleum and coal, will greatly slow down the pace of transportation and circulation of physical goods within globalization. After all, without the support of energy and resources, globalization and the global village may fall back and become a beautiful vision once again. This means that the scope of physical goods' transportation and circulation will gradually shrink, and localized production and manufacturing will likely also once again become mainstream.

At this point, the traditional concept of globalization will evolve into a combination of digital modeling globalization and manufacturing localization. A "supplier," or a designer, will provide the design program with a product's printing solution on a computer and will store it as a digital model.

Then, they could sell the design program to the "demander" in any part of the world, and the transaction could be done via the Internet. After receiving the design program, the "demander" can print out the physical product in their location via the 4D printer, thus realizing the localized "production and manufacturing" of the product.

Such a convenient way of transaction and product manufacture sounds like it is from sci-fi. In fact, a sharing website called Thingiverse, where designers can upload their digital works for others to download or print out, has become a reality, and there are already many people who regularly share files, pictures, and videos online.

The magic of 4D printing goes far beyond this. Relying on the changes in the parameter settings of the time dimension in 4D printing, the locally printed products will also be able to make adaptive changes according to the physical environment of their locations. This could be the so-called "tailor-made" aspect of the design, and it could be difficult to achieve when products are produced in a unified production base within the traditional model of globalization.

In the era of 4D printing, printing objects will realize the storage and transmission of digital information in the physical world through computers and the Internet. The problem of long-distance physical goods transportation constrained by the lack of energy and resources will also be resolved. As such, the waste of energy and resources will be minimized to improve the efficiency of the transmission of goods.

7

THE MAKER

There is an old saying, "There is but one secret to success: never give up." The saying is also incisive in the development of personal 4D printers because people who are pushing 4D printing out of 3D printing models are called the "Makers."

SECTION 1 | NEW ERA OF THE MAKERS

The Makers refer to a group of people who like or enjoy creativity. They pursue the realization of their creativity through DIY projects, which means they like to make things by themselves and turn various ideas into real products. They are "super users" of software companies, and also a bit like "hackers," they modify the software and improve technology according to their own will and expectations. Besides, when the Makers advocate personalized customization, they also advocate open-source sharing. This reflects self-confidence, whereby they believe that even when all technical details are free and open, no one in the world could do better than them in this small field.

Even though the "Makers" are just sprouting in China, they have a long history of development abroad. Steve Jobs is one of the pioneers in the world of the Makers. He's been obsessed with innovation since he was a child, and he once tried to change the pulse frequency of the telephone to make free calls. His persistent pursuit of innovation makes Apple what it is today. Furthermore, the diversity and sharing in the Makers' culture make it inclusive and full of affinity, allowing more people to join in. Star Simpson, majoring in Electrical Engineering and Computer Science at MIT, was addicted to open-source hardware. In her opinion, there is no difference between low tech and high-tech, and they both are tools used to bring our ideas to life.

With the development of the Makers' culture, a lot of novel technologies and products were born, and the traits of the Makers came to the fore. The first is that they must do the work themselves. The second is that they must share the novel stuff they've made, which means sharing thorough design drawings and source codes with everyone. The third is that they have to take advantage of network communication to work with like-minded people to discuss and improve products' creativity positioning, technical difficulties, financing methods, profit models, and marketing methods, among other factors. It is their enthusiasm and selflessness that break the technological blockade of large factories, large companies, and multinational corporations and lay a solid foundation for the arrival of a new era of "personal intelligent production," "family intelligent production," and "network community intelligent production."

SECTION 2 | 4D PRINTING FOR CREATIVITY

While people are still exploring B2C and discussing C2B, the applications of 4D printing technology based on 3D printers have been weaving another technological blueprint for the future world—a unique and personalized world showing independent creativity. In this world, all people create, and everyone is an artist. Everyone's items are unique and exclusive to their possessions only. Every household can set up its factories to break large traditional manufacturing enterprises into pieces. What's more, the 4D printed personalized products will transform into infinite possibilities with the changes of the time dimension and living space. Let's imagine that the items we use can transform with the changes in the existing space and triggering medium. The clothes we wear can transform with the growth of our body and the changes in our composure. Even the organs in our body can constantly repair, improve, and upgrade as the living conditions change. All these ideas are possible with the advent and development of 4D printing, which will make our future world magical and creative.

3D printing has led product customization to cross the high-cost threshold, and 4D printing has further lowered down the cost for the personal customization of products. It has enabled it to move forward without cost constraints and reach a rapid upgrade.

The basic requirement for the popularization of low-cost product customization is there the sharing of the design program of digital products.

Once, we praised design software that shaped our world because there was a computer design file behind almost every architectural model, every product prototype, and every finished product. This includes the chair you are sitting on now, the book you are holding in your hands, the computer on your desk, the car you drive, the house you live in, and even the buttons on your shirt. All these things were digitalized before they were produced. In short, the design file is the language of modern engineering.

Before and even after all the products are produced, design files are often not made public and kept secret. Because they are not only the fruits of designers' labor but also the expression of their wisdom. They are always "protected" under "intellectual property."

However, with the promotion of 3D printers and the development of 4D printing technology, the Makers have come up with more design programs through their creativity. Even more and more individual Makers and Maker groups are emerging, thereby more and more creativity and products are coming to life. Because of Makers' self-confidence and their advocacy of resource sharing, every "eager" heart in the world can enjoy the "welfare" of design documents. As long as people have good intentions, they can easily obtain the original programming codes of product design released by the Makers from the relevant media or network information platforms. Thus, they will take part in technical communication and learning interaction at the same time. Furthermore, everyone can partially modify, adjust, and redesign the products according to their preference based on the original design by Makers. After fine-tuning some of the code and programs, new product design codes can be fully incorporated with your aesthetics and preferences. Then, the product becomes your unique and exclusive "personally customized" product in the real-world. By promoting such practice, as long as someone wishes to do so, they can create a personal-customized product on the fertile resources established by the Makers.

A 4D printing movement of "everyone's creative ideas and everyone's designs" has been brewing and fermenting, and it has slowly set off the prelude to a new production model for the manufacturing industry.

SECTION 3 | 4D PRINTING FOR CIVILIANS

As the saying goes, "You can know the distance from the near, the more from the less, and the salient from the subtle." We can see that from the efforts of the Makers, soon, right after 3D printing, 4D printing will enter all aspects of our future life and work.

At the second White House Science Fair on February 7, 2012, a group of Makers was invited to the White House State Dining Room to demonstrate the innovations to the public. Joey Hudy, a 14-year-old Maker, with the help of US President Barack Obama, fired the "extreme marshmallow cannon" that he invented. In 2012, in Amsterdam, the capital of the Netherlands, a modern art exhibition focusing on the latest achievements of 3D printing technology kicked off, which made us fully feel the infinite power of the public's creativity. As a great man once said, "The people, only the people are the driving force to create world history."

With the combination of 3D printing and 4D intelligent data, the threshold of product design and manufacturing almost becomes zero. By facilitating ordinary users to easily group themselves as Makers, it will be possible to kickstart the miracle of the new Industrial Revolution and move the expensive 3D printers from high-end factories into thousands of households. Additionally, the development of digitalization will usher in the first full-blown period of "personal hero opportunity" in the history of human society because we are all born Makers since the very first clay figurine we made in our lives as a child.

If things go on like this, people will continue to upgrade their fantastic ideas and apply the tools that are already advancing in our daily lives, thus promoting the overturning of the existing technologies and starting the revolutionary waves set off by the new technologies. The promotion of 3D printers as well as 4D tech, which is an upgrade on the technical level, will help to move humanity past the castle of the "high-tech" and into the homes of ordinary people. All of this will be pushed by the interest and efforts of the Makers.

SECTION 4 | ENTREPRENEURSHIP STARTED BY A PRINTER

The popularity of 4D printing has turned everyone into an inventor full of creativity. With the promotion and applications of 3D printers and 4D printing technology, we can not only print the needed personalized items at home but also easily start a business as a manufacturer of 4D printed products. As long as there are needs, whether models, parts, or even food or cars, we can print creative and personalized products for them through a click of a mouse.

Similar to the agricultural era, the original acre of land was self-sufficient to support a family. But now, the limited cost of a printer allows everyone to easily set up their own "manufacturing" factory and embark on a "production" entrepreneurial road. Such an enterprise will minimize the costs of labor, materials, and finance for 4D printing manufacturers. This means that an individual plus a printer (including materials) will be all that's needed. For families living in a neighborhood, they could divide the work to meet their different needs. All these can be called manufacturing by all. Different from the ancient arable and weaving, today we see a prosperous scene of a Chinese society where every household is producing and manufacturing goods to fulfill the material needs of society, science, and technology. As we are relying on the advancement of science and technology, we are not using hammers and hoes, but the high-tech 4D printing equipment and memorable polymer materials. What we are producing is not just grains, but various items that meet the needs and aspects of our life and work.

Clothing. In the 4D-printing manufacturing process, households in charge of clothing production and manufacturing will create a real "personal customization" history of human clothing as well as enter a new era of "one piece per person." As such, only one piece of clothing will have the ability to self-transform into various forms. First, it will transform into different shapes and colors according to the wearer's needs on different occasions. Second, based on the contact with the human body, it will also adjust its thickness and materials with the change of the human body's temperature to meet the wearer's needs of wearing the same piece during all seasons. In addition, it will even be pre-programmed with the needs of the human body's skeletal growth, growing together with the wearer. Every piece of such clothing will truly realize the "personal customization" for the clothing manufacturing industry.

Food. The 4D printing of food will change the bad habits of people's traditional diet, and we will be able to say goodbye to "junk food." First, people will be able to keep the goods and discard the rest in the production of raw materials, which largely breaks the current predicament of "illness entering through the mouth." Second, the enhanced matching between food ingredients and human body functions will avoid the negative impact on the human body and the food waste caused by "reckless eating." Thereby, it will truly realize the "personal customization" of diet.

Housing. Regarding the 4D printed houses, we have already introduced the houses fixed in a certain place in other chapters. Here, we are going to dream freely about mobile residences, which will play an important role in the traveling life of people pursuing the future. First, a 4D printed mobile residence will be set with the user's personal preferences. Hence, its transformed form will meet the user's individual preferences to the greatest extent. Under the action of the medium, it will transform into a new-style tent for users preferring tent form and transform into a recreational vehicle for users preferring recreational vehicle form. Second, the 4D printing raw materials used in the manufacture of mobile residences will have greater shrinkage and curling capabilities. They will also be able to recover under the action of the medium to achieve multi-repetitive use.

Transport. As a representative of transport, we have introduced 4D printed BMW in a separate chapter. In the future society, more convenient transportation means perhaps just a pair of shoes on our feet, just like the Wind Fire Wheels on Nezha's feet.* Maybe a chair under our butt could be a new driverless convertible car, or even a hat or an umbrella over our head, could take us flying with just a single click. All these incredible types of equipment will come along with the continuous advancement and development of 4D printing technology, the development of raw materials performance, and the program coding presets. Plus, the popularization and applications of science and technology development to all these factors will also realize these dreams.

SECTION 5 | THE SOURCE OF POWER FOR 4D PRINTING

In the 21st century, each of us wants to use unique products, just as each of us is unique. This makes us believe that the market is full of endless business opportunities, and personally customized products are lucrative. Makers are forming a new industrial organization model, which is an interest and project-oriented driven company on a smaller scale with tendencies to be virtual and informal. The Makers team up and regroup their operations. The team

---

* The Wind Fire Wheel is one of the well-known magic weapons in Chinese mythology. The sound of wind and fire can be heard when the two wheels rotate. The wheels belong to Nezha who is an ancient Chinese mythical creature and is a legend of the immortals. While Nezha's wearing them, he can freely fly.

members in the company are far fewer than the traditional large companies, but their innovation ability is much stronger. Makers not only perform well in niche markets but also frequently set off huge waves in mass markets. Thus, the existence and development of the Makers will become the lubricant and accelerator for the development of 4D printing.

If we go back in time, if it weren't for Steve Jobs, it's hard to imagine that Apple would have grown into the world's most valuable company. When Jobs left, Apple weakened and near bankruptcy. When Jobs came back, Apple prospered. First of all, it took an iPod music player to make Sony's Walkman, which had dominated the world for many years, extinct. Then, it used the iPhone to make Nokia, the world's No. 1 mobile phone company, announced that it would never make mobile phones again. You have to know that Apple is only responsible for designing mobile phones, and they have never manufactured any of them but outsourced to FOXCONN, owned by Terry Gou, for their manufacture. Therefore, it can be said that Apple defeated Nokia with almost bare hands. In the era of the new Industrial Revolution, through collaboration and the sharing of inexhaustible network knowledge, the power of individual creativity will be magnified many times. Sometimes a single Maker is enough to overthrow a traditional industrial empire, just as easily as taking something from a pocket.

There is another US Maker named Elon Musk. He is considered to be the role model prototype of the movie *Iron Man*. In 2002, he sold PayPal (the world's largest online payment platform, similar to China's Alipay), co-founded by himself and his partner, to the world's largest online store, eBay, for US$1.5 billion. After that, he founded a company, Tesla, and produced the first electric vehicle in the world that could accelerate from 0 to 100 km/h in four seconds. It was successfully mass-produced. In 2010, the Falcon 9 rocket launched by another company he founded, SpaceX, successfully parachuted the "Dragon capsule" back to earth. This is the first time in the history of the world that a private company launched a spacecraft into space and managed to receive it back. The entire aerospace industry was shaken. The standard launch cost of the Falcon 9 rocket is US$54 million, compared to the cost for the Aeronautics and Space Administration (NASA), which is US$435 million. NASA subsequently announced that they would not build any new space shuttles in the US after the existing ones retired in 2011. Instead, they would entrust private companies like SpaceX to deliver supplies to the International Space Station. They signed a contract worth US$1.6 billion with SpaceX. It can be said that in terms of rocket manufacturing and launch, Musk defeated NASA completely by himself. Moreover, Musk plans to invent a reusable rocket, hoping to realize the dream of human migration to Mars within the next ten years.

It is the Makers' DIY abilities that keep opening up the sources of human innovation. Today, when 3D printing and 4D intelligent digital tools become the "nunchucks" in the hands of the new generation of Makers, the world will be changed. With the birth of the 4D printing concept and the development of its relevant technologies, Makers will, as always, receive more glory in the era of technology.

SECTION 6 | COMBINATION OF 4D PRINTING AND MAKERS

The advantage of the traditional manufacturing industry is the standardization of assembly lines and their large-scale production, which is the most effective way for development and profitability during a commodities shortage. However, in today's world, where the production means and living materials are becoming saturated, the needs of consumers are turning toward cultural and spiritual thought.

In this process, new production means will be appreciated. For example, the emergence of new production means, such as intellectualization and 4D printing, conforms to the stage of our historical development. It creates the appearance of inevitable technologies under the development of science and technology. On the other hand, the change in consumers' demands for commodities leads to the thinking and emergence of new technologies.

The emergence and popularity of the Makers is not a product of the era of industrialized mass production. In my opinion, it is more about thinking, innovation, and exploration of the niche and vertical fields, which are also the core of Makers' survival and development. In traditional industrial thinking, there is a proportional relationship between volume, cost, and profit. The reason why it cannot be broken is that the traditional industrialized manufacturing methods are usually based on the production process of molds and standard assembly lines.

That is to say when traditional industrialized manufacturing methods are used for small batch or personalized product manufacturing, it is difficult for the input and output to achieve profitability. Therefore, industrialization is bound to rely on new production means and production methods to solve the problem of traditional manufacturing methods. This is also a core factor of why 3D printing is concerned in this movement and looks promising.

The impact of the revolution brought about by 3D printing in the manufacturing industry is far-reaching. It is satisfied with not only the production methods of "personal customization" but also the low-cost rapid manufacturing. Nevertheless, the impact brought by 4D printing will probably be even more thorough.

If 3D printed objects, such as furniture, take up more space in transportation and bring inconvenience during assembly, 4D printing can completely solve these problems. We only need to input the transformation memories into the printing material and models. Let's say a cup has water as its triggering medium; it can then automatically transform and adjust according to the amount of water stored in it. This means it won't be inconvenient for us if we want to drink a small amount of water from a big cup.

Of course, it is not limited to these products only. For Makers, 4D printing solves the problems of personalized production and manufacturing. Meanwhile, it puts creativity into a pair of wings, which can self-transform and self-assemble. This pair of wings will be able to realize creativity, give more creative intelligence, and allow more creative ideas to happen within the products themselves in the coming era where everyone will be a Maker, and everyone will be creative.

8

ETHICS

A seemingly ordinary strand was gently placed in the water, and a miracle happened over time. The strand slowly moved, rotated, changed, and went through twists and folds to eventually form a regular tetrahedron. The amazing part was that there was no external equipment involved at all. At the TED conference held in Long Beach, California, the US, Skylar Tibbits, a lecturer from the Department of Architecture, MIT, demonstrated the fantastic 4D printing technology. When the movie-stunt effect, self-transformation, was truly presented in front of people, it was no surprise that it left the world with wonder and admiration.

When Skylar Tibbits was the first person to reveal the topic of 4D printing technology that met the needs of humankind and even the planet in this unique way, issues related to intellectual property and laws in this field were raised. The same issues were entangling the development of 3D printing technology, but it was still a big source of concern for stakeholders in the relevant industries in the development of 4D printing technology.

SECTION 1 | LEGAL RISKS AND CRISES

Presently, 4D printing is still in the early conceptual stages, and its relevant service markets are also in the infancy stages. So, there are very few people involved in this open ultra-frontier field, and the commercial atmosphere thus is yet to flourish to attract widespread attention from the public. However, as an emerging technology that will change the rules of its respective industries, its subsequent potential influence is unpredictable. Just like our personal computer, in only ten years or so, it has fundamentally shaken our legal system. And today, AI, represented by ChatGPT, is impacting our society like never before.

So what new legal challenges will the 4D printing technology, which has already been pushed out of the water, encounter in the near future, and what new forms of consumer security and criminal activity?

Previously, making counterfeit currency was technically challenging. However, with the development of computers, the thresholds for manufacturing counterfeit currency have all been lowered significantly. Before 1995, less than one out of 100 counterfeit currencies were made using computers and laser printers, and just five years later, in 2000, nearly half of the counterfeit currency was designed online and then printed on high-end color printers. Computer design software, color laser printers, and toner technology have made it easier to fake counterfeit currency. The economic damage caused by counterfeit currency is self-evident. Through 4D printing, it can not only easily make counterfeit goods but also make guns, ammunition, other weapons, and even drugs without the country's permission. The risk of personal injuries to the people is chilling.

In 2012, the first small-scale ethical conflict in the field of 3D printing broke out on a file-sharing website called Thingiverse.com. Meanwhile, such conflicts will also appear in relevant fields after the emergence of 4D printing. According to the report, a user uploaded to this website a design file for printing a rifle part by using plastic as raw material on a consumer 3D printer. The particular rifle part was the only one that required a background check from the user in the gun manufacturing process, but the design file made it possible for people to own a rifle without a license. In other words, parts that are made through 3D printing technology can circumvent gun control laws.

In the future era of 4D printing, what the "finished product" or the printed "semi-finished product" will transform into under different medium conditions may be unknown even to the person who prints it. Unlike the appearance of a gun, it may look more like a shoe or a brush at first. This further increases the difficulty of the relevant authorities to monitor and control printed objects. Moreover, through the application of 4D printing technology, the manufacturing costs of guns will be significantly reduced. As the performance of the printed rifle was summarized by the netizens, the 3D printed plastic part was so powerful that it could fire 200 rounds. To complete this, there was no need for special equipment, just a long service life Stratasys printer could print out the part, and the raw material used was just ordinary commercial-grade resin costing US$30 only.

Other than firearms, the magical 3D and 4D printers will easily print out drugs, medicine, and many other unimaginable items. Traditionally, the production of pharmaceuticals requires meticulous research and validation to ensure their safety and efficacy. However, with the potential for individuals to possess 3D and 4D printers, people can design and produce their own drugs without the strict regulation of the traditional pharmaceutical industry. The production of drugs without professional knowledge and regulation could lead to adverse drug effects and side effects, posing potential risks to individual and societal health.

The misuse of 4D printing could also enable the production of more potent chemicals, possibly including dangerous drugs and chemical substances. This presents a significant challenge for regulatory agencies, as traditional regulatory means may not effectively monitor the variety of compounds produced individually. This not only increases the risk of societal drug and chemical abuse but could also lead to more criminal activities.

In the US, the "War on Drugs" has unfortunately failed. Prisons are full of non-violent criminals, and high taxes are being spent on arresting drug addicts rather than the lower cost and more effective drug rehabilitation programs. The number of deaths from overdoses of prescription drugs in developed countries has soared. Imagine if someone makes a batch of chemicals that can change people's moods by using the printed reaction vessel and then spreads these products freely. What a catastrophe that would be!

It can be said that the more powerful the technology, the more novel and profound the potential for misuse. This will become a nightmare for regulatory bodies. Therefore, in addressing this issue, regulatory bodies need to develop more flexible and innovative regulatory strategies to keep pace with the rapid development of technology. At the same time, society also needs to be involved in the regulatory process, working together to maintain public safety and health.

SECTION 2 | THE LOST INTELLECTUAL PROPERTY

The 3D printing technology is incredible. It allows manufacturers requiring complicated craftsmanship under traditional technical conditions to complete it with a simple click of a mouse, and a variety of crafts, toys, clothing, and violins could be easily printed out. 4D printing is even more fantastic, making the "Transformers" in science fiction films and the "*Jingu Bang*" in the myths a reality. The printed objects could transform into anything you wish.

Obviously, the impact of 4D printing on the future of the world will be huge. Among all these impacts, the protection of intellectual property has first entered people's field of vision and has become one of their focus and concerns.

In fact, in recent years, along with the rapid development of 3D printing, disputes about intellectual property rights have risen and fallen. A British game company issued a cease-and-desist "order" to a 3D printing vendor for printing physical models of the character from the company's popular desktop game "Warhammer." A Dutch designer sent an "off-line notice" to Thingiverse, which is an online database of design codes for 3D printers that allows anyone to download and share designers' designs for free. The controversial Swedish file-sharing website, The Pirate Bay, which has been the target of lawsuits around the world, recently announced that it was embarking on a new service to share 3D printing designs. The US cable network HBO sent Fernando

Sosa a "cease-desist letter" asking him to stop selling 3D printed iPhone docks mimicking the Iron Throne in the HBO series *Game of Thrones*.

It is apparent that the legal protection of intellectual property of 3D printing is still an issue that needs to be considered and urgently resolved. Yet, for the intellectual property of the coming 4D printing, it's more like a lost lamb—where will it be led?

We will shudder if we follow 3D printing's thinking on intellectual property issues because 4D printed items could change shape at any time according to the changes in the set time dimension and stimuli. Let's assume that a designer 4D prints a ring, which is just a representation of a semi-finished product, and ultimately it may be a rose, a gun, or even drugs. Such changes in 4D printing are not just a dream or hypothesis. Then, without a setting in time, people won't be able to identify what the printed item will ultimately be; thereby, they won't be able to evaluate the ownership of the intellectual property. However, even though the time will be preset, it will be meaningless for 4D printing as people could avoid the preset time while designing.

In the face of such embarrassment, can we boldly and tolerantly assume that 4D printing is more like children playing with building blocks? Whatever gets built is what it is. If this is the case, who can say that what a person has built must have a patent and intellectual property protection, while others cannot build the same or similar things? In fact, given this scenario, the issue of patents and intellectual property will be shifted to building blocks instead of how they are built. In the same way, the intellectual property of 4D printing can create a new situation by shifting these concerns. It's worth waiting to see what will happen.

SECTION 3 | CHALLENGES TO INTELLECTUAL PROPERTY PROTECTION

Intellectual property protection is becoming more and more important in the period of the rapid development of science and technology. Within the current background of mass entrepreneurship and mass innovation, only a perfect system of intellectual property protection can bring lasting power to innovation and economic development. With the conditions of traditional manufacturing technology, including personalized customization based on Industry 4.0 technology, although the production cost of personalized customization has decreased compared to traditional manufacturing technology, the production cost of personalized customization for the individual user is still high compared with 4D printing technology.

This means that with 4D printing technology, production costs will be greatly reduced, especially with the increasing maturity and popularization of desktop printers. This will fundamentally hinder or shake the interests of traditional manufacturers. It is foreseeable that in the era of 4D printing, few people will be willing to pay high prices for branded products but rather buy low-cost raw materials and print products at home with their 4D printers. As we can see, 4D printing poses new challenges to the existing fair use system of intellectual property, as well as the monitoring and evidence collection systems. In particular, the difficulty of intellectual property protection caused by 4D printing's capabilities of self-transformation and self-assembly triggered by adding the time dimension will be far beyond those of 3D printing.

Currently, we still lack the related intellectual property protection systems for 3D printing technology. Compared with 4D printing technology, intellectual property protection for 3D printing is relatively simple and it focuses more on how we could solve the infringement behaviors of "imitation and copies." However, 4D printing technology creates a further subversion and transformation from 3D printing. The addition of the time dimension makes the printed objects no longer a simple copy, but it integrates factors of self-transformation and self-assembly, which pose a challenge to traditional intellectual property protection.

The current problem is that the existing laws of intellectual property protection, blank on both, 3D printing and 4D printing. This is not only a problem in China but also a common challenge faced by countries around the world. Now, the problems faced by 4D printing are that as technology advances, various requirements of intellectual property protection will be inevitably put forward, and there will therefore be mainly three challenges. These are described below.

The product form of public transactions. Due to the self-transformation and self-assembly capabilities of the 4D printed products, the intellectual property protection of commodity transactions is more complicated. One example is a home appliance. The 4D printed home appliance is a copy of a certain brand, and while going through the evidence collection process conducted by the regulatory authority, it can self-transform and self-assemble into another form which is non-infringing. Let's say a refrigerator is copied from product A. In the course of a display or an evidence collection, the refrigerator could self-transform and self-assemble into a form with different sizes and shapes that have been preset in the changeable design model. This makes it somewhat more difficult to obtain evidence for intellectual property protection.

The gray edge of customization transactions. Just like some business models and services were derived from 3D printing, there will be some professional printing platforms that will provide services to customers derived from 4D printing, as well. However, the forms presented by the platforms will be more diverse and varied; perhaps they will follow a crowdfunding model or belong to a few people with a common interest in serving a community or a group. With the help of software synthesis technology of 4D printing, when users see a product, at any time and place, they can take a photo of the product by using a cellphone or a wearable device and send the photo to the selected customization platform. Then, the platform providing a 4D printing service will synthesize the photo with the aid of software and print out the corresponding product for the user. Such service derived from customization will push intellectual property protection to the gray edge.

Copyright challenges for personal printing. When desktop 4D printers enter households, especially for some smaller products, such as pots and pans, teacups, and home decorations, among others, users could then complete the printing at home depending on their preferences. When users surf the Internet and find a product they like or a creative design they are interested in, they can simply download the image to the corresponding 4D printing software for 3D synthesis and then send it to their 4D printer to print out the product. Not only that, but users could also print a variety of different designs into one and trigger the printed object to transform to fulfill their needs when needed. Thus, we can imagine that the difficulty of evidence collection of infringement could occur in the process.

Due to the misalignment of law updates and the development of science and technology, the loopholes caused in the legal perspective of intellectual property protection have been presented. Therefore, the questions of how we can adapt to the requirements of new technology development, find a balance between new technologies and innovations, and establish a protection system for 4D printing's intellectual property will soon be the possible common concerns facing industry experts and regulatory authorities. As far as I am concerned, we should explore and establish relevant laws and regulations without delay to better promote the healthy development of the industry and minimize the negative impact brought about by the new technology revolution.

SECTION 4 | COPYRIGHT

Intellectual property is a large legal concept that includes copyright, trademark rights, and patent rights, among which, whether we are talking about 4D printing or 3D printing, copyright is the most closely concerned concept.

Copyright is primarily applicable to creative works. It gives the original copyright owner exclusive rights within a certain period of time to arrange their work (the specific period varies from country to country, but it is typically the life of the copyright owner plus 70 years after their death). If others or other organizations want to reproduce, modify, sell, rent, or publicly present the work, they must obtain permission to do so from the copyright owner.

The key to copyright is the price of the artwork and not its usage or practicability. Michael Weinberg, a lawyer and a 3D printing expert, once explained that "making things" itself cannot be granted copyright. The scope of practicality is wider. For example, clothes are useful, and you can have the copyright of clothes' design but not for the concept of tailoring.

Industry experts predict that the most challenging issue in copyright disputes in the future will be the recognition of the copyright of "derivative works." According to Copyright Law, a derivative work is an expressive creation based on the original work. Translation is a derivative work, an imitation of a popular song is also a derivative work. If you edit a design file of an existing work and turn it into a new one, under what circumstances are you creating a derivative work? Under what circumstances are you creating a brand-new work (a brand-new work means you are the copyright owner)? Or under what circumstances are you deliberately taking the fruits of others' labor?

4D printing can be regarded as a kind of "derivative work" in some ways. The concept of copying "works of applied art," and illegal copying are exactly what copyright must prohibit, particularly for the copyright protection of "product appearance and structure" involved in 4D printing. The current Copyright Law in China is not sufficient to protect the copyright of "product appearance and structure." Those few "product appearance and structures" with artistic value can be protected as artworks, while it is difficult to protect most ordinary "product appearance and structures" under the Copyright Law. The Copyright Law of China is facing a third revision, which draws on the relevant provisions of the Berne Convention. It is adding provisions on copyright protection for "works of applied art," and formally including the "product appearance and structure" into copyright protection.

In the process of 4D printing, if the "copying" is done without the author's authorization, it is considered an infringement. However, while judging the infringement of 4D printing, the method of printing also needs to be looked at. So far, no matter if we are discussing 4D printing (whereby the semi-finished product produced by the printer with the time dimension transforms into its final form triggered by a specific medium) or 3D printing that directly prints a product into a form, both processes involve the following three methods. First, the method of printing from three-dimensional to three-dimensional objects, that is the printing of a three-dimensional model from the computer to a physical object. Second, from text to three-dimensional, that is, entering textual descriptions on the computer, such as rectangle, height 18 cm, width 20 cm, red in color, etc., to print out a semi-finished object or the corresponding object. Third, printing from a flat to a three-dimensional model, that is, printing the flat image on the computer through the printing program into a three-dimensional object which will transform into its final shape by triggering the medium.

The first method of three-dimensional to three-dimensional printing is a typical copying behavior. For flat-to-flat or three-dimensional to three-dimensional printing, both methods belong to the typical copying when it comes to Copyright Law. Even if it is a reduction printing, enlarging printing, or other methods of changing the proportions, it won't affect the fact that it would be considered copying. Therefore, copying without the author's permission will constitute an infringement.

The second method of text-to-three-dimensional printing is usually not considered as copying under the Copyright Law. The Copyright Law protects "expression," and text and three-dimensional printing are two different forms of expression, so it doesn't involve copying. For the same reason, this printing method for 4D printing and 3D printing generally won't be affected by the infringement.

Is 4D printing from a flat model to a three-dimensional model considered copying? China's Copyright Law is silent on this topic, but it is quite controversial in practice. In the "Fudan Kaiyuan case" in 2006, the defendant converted the form of the Chinese zodiac from flat cartoon images to three-dimensional money boxes without legal authorization, and the court held that this was an infringement of the plaintiff's copyright. However, in the "Motorola copyright case," the court held that Motorola's act of producing printing circuit boards following their design drawings was an act of industrial production and thus did not belong to the act of copying under the Copyright Law. The court's judgment on whether the same "flat to three-dimensional" method constituted copying was completely different. When we refer to the provisions of the Berne Convention on "copying," which includes copying in any form and any way, such open-ended phrasing raises a higher standard for copyright protection. More importantly, in the era of 3D and 4D printing, such "copying" methods will inevitably multiply. So, it is necessary to clearly state the "copying" methods in legislation to protect the legitimate rights and interests of copyright owners.

SECTION 5 | TRADEMARK RIGHTS

Much like copyright, trademarks are also considered intellectual property. People usually use "brand names" or "logos" to define a trademark. The information it conveys is that it is a registered mark or "trade dress" that shows consumers that the product is produced or provided by a specific manufacturer. The original purpose of trademarks was to protect the interests of consumers. Gradually, trademarks have since become a marketing tool, and those highly recognizable trademarks can be worth billions of dollars.

In the development of 4D printing, it is likely, that trademark infringement will also be involved. For example, users may print out the trademark together with the product, such as NIKE shoes. A printed pair of NIKE shoes will usually carry the NIKE trademark, and if this printing does not obtain authorization from NIKE, it will very likely constitute an infringement. Other than that, for the rest of the printing methods, if one is simply printing the product itself, that is, only printing out NIKE shoes without the NIKE logo. Generally, this should not involve trademark infringement but may result in other intellectual property infringement.

Much the same, 4D printing technology also opens up the door of convenience for the production and manufacturing of trademarks. Although this is a relatively independent printing behavior, even if the user only prints out the NIKE trademark without printing the shoes, this printing may also lead to trademark infringement. This is because according to China's Trademark Law, the act of manufacturing and selling a trademark without authorization from the trademark owner would be considered a trademark infringement.

While 4D printing, users can also print according to their ideas, such as by following a printing method like only printing NIKE shoes (with no trademark), but at the same time, later printing the PUMA trademark on the shoes. This means that the NIKE shoes will carry a PUMA trademark. Would we consider this an infringement? Whose trademark right does it infringe? First of all, if the user doesn't have authorization from PUMA, the unauthorized use of the trademark may constitute its infringement. Then, what about NIKE? Does such a printing method constitute a trademark infringement for NIKE? In China's Trademark Law, there is a provision called "reverse infringement," which means if you replace the NIKE trademark and put the products on the market without permission, it still belongs to trademark infringement. So, if a user prints out products by using the above method and puts them on the market, their behavior infringes the NIKE trademark as well.

SECTION 6 | PATENT

The word "patent" is derived from the Latin "patere," which means "to lay open." According to Wikipedia, a patent is "a form of intellectual property that gives its owner the legal right to exclude others from making, using, or selling an invention for a limited period of years in exchange for publishing an enabling public disclosure of the invention. Like any other property right, it may be sold, licensed, mortgaged, assigned or transferred, given away, or simply abandoned."

The core technology of 3D printing, "laser sintering," has a patent. In January 2014, the patent protection period expired, which allowed the 3D printing patent to enter the public domain, largely promoting the rapid advancement of 3D printing. With the advent of a large number of 3D printers, the manufacturing revolution is now ready to take off, and in this process, the issue of patent infringement will undoubtedly attract more attention.

According to China's Patent Law, the types of patents can be roughly divided into three categories: inventions, utility models, and designs. Invention and utility model patents focus on the internal structure and innovation of a product, while the design patent focuses more on the appearance and color of a product. Unlike the existing flat printing, either 3D printing or the further development of the new 4D printing involves both the appearance and the internal structure of a product. Hence, the three categories of patents are closely related to each other, which undoubtedly requires a user to be familiar with the patent protection of the product they want to print in advance in order to prevent infringement of others' patents.

The development of 3D printing and 4D printing may also promote the commercial use of expired patents. According to the provisions of China's Patent Law, patents have a protection period. The protection period for invention patents is 20 years, and it is ten years for utility model patents and design patents. If the protection period is exceeded, the patent will enter the public domain, and everyone can use it for free. Many of the expired patents are full of creativity in appearance and structure, but they are limited by technical limitations and are difficult to realize. Therefore, they are left unused. A US lawyer has discovered a lot of interesting and useful designs out of the expired patents and set up a special section on the Internet for users to download the expired patented 3D design drawings for their own printing. This will undoubtedly be used in the field of 4D printing.

SECTION 7 | "FAIR USE"

When discussing 4D printing and intellectual property infringement, we must also look at the issue of "fair use." So, what kind of behavior is called "fair use?" As regulated by China's relevant laws, "fair use" in the Copyright Law refers to the use of a work in a certain way without the consent of the copyright owner or without the payment of remuneration to the copyright owner. In the intellectual property system, acts of copying a small amount of published work for personal use only, such as for self-study, research, appreciation, use in the classroom for teaching twenty teaching research, among others, are regarded as "fair use." For these instances, the permission of the copyright owner is not necessary, and the use is excluded from the scope of infringement.

What if a user simply 4D prints personal consumer goods and does not use them commercially? Does their behavior constitute infringement? Under the existing Intellectual Property Law, such behavior is generally regarded as "fair use" rather than infringement.

If this is the case, we can imagine that in the future society, few people will spend a lot buying branded products. Instead, people will be more willing to spend less on buying low-cost raw materials to print out the needed products at home. The result of "fair use" by many people will lead to the bankruptcy of merchants. With no surprise, 4D printing also challenges the current fair use system.

Under the conditions of traditional technology, the production cost of the individual user is relatively high. Therefore, production for "research, appreciation or personal use" is extremely limited in scale and does not fundamentally hinder the interests of merchants, so much so that there is no problem with the behavior being considered fair use. But, under the conditions of 4D printing, the production cost will be greatly reduced, which will make the idea of "fair use" fundamentally hinder the interests of merchants. By then, revisions of "fair use" clauses will be needed, and a contest between the public and patent owners will be inevitable.

Even now, the adoption of technical means to prevent "fair use" from damaging the interests of merchants has been put on the agenda. The US Patent and Trademark Office recently introduced a "production control system" for copyright protection of 3D printing. Under the system, before any 3D printing-related device performs a printing task, it must compare the printed model with the data in the system database. If there is a large proportion of matching, the 3D printing task cannot be performed. Both the Digital Millennium Copyright Act of the US and the Copyright Law of China support copyright protection through technical protection measures, and behaviors that breach technological protection measures are regarded as illegal acts.

SECTION 8 | PROTECTION OF CONSUMERS

Along with the maturity and popularity of 4D printing technology, when more and more people use 4D printing to produce the items they need, potential risks will continue to grow and spread while they enjoy personalization and convenience. These risks go beyond the traditional legal framework of consumer safety. Under the existing legal mechanisms, first, no one can ensure that consumers will not be able to buy products that hurt them. Second, once harm occurs, there is no way to attribute responsibilities between the parties involved.

According to China's Tort Law, if a party can prove that they have made reasonable efforts to prevent and minimize the risk of product failure in the context of fair use, the party's responsibility ends. A simple example is tire manufacturers. If the tires they produced exploded every time the speed exceeded 120 kilometers per hour, they would be held liable even if it was illegal to speed at the time. This is because speeding is still common in everyday life. However, if the buyer of the tire drives into the roadside and the tire explodes, the tire manufacturer will not be held liable.

Standards help to set clear boundaries of responsibility. In the early days of steam engines, boilers often exploded, frequently causing damage. In response to this situation, insurance companies have set explicit production standards to divide responsibilities. Finally, a series of standards have been established, such as standards of material thickness, safety margins, and pressure relief valves, to ensure that filters can withstand the lowest steam pressure requirements. Boilers that do not meet these standards will not be eligible for insurance compensation.

Such a legal mechanism with a clear allocation of authorities and responsibilities is also worth referencing during the age of 4D printing. In contrast to the current situation when 3D printing is popular, an inevitable scenario can be that a car hobbyist could be killed in a car crash due to the disengagement of the purchased 3D printed steering wheel while driving at high speed. When the defense lawyer is defending the victim's family, they could be confused as to where they could attribute the responsibility. Should be the person who created the wrong design file? The person who 3D printed and sold the part? The websites that advertised the part? Or the manufacturer who installed the part on the car, the manufacturer of the 3D printer, or the raw materials supplier?

In the absence of clear and standard boundaries of responsibility, no right of recourse would be imposed on manufacturers of 3D and 4D printed products and as such, they could irresponsibly seek to maximize profits. Such a situation may be familiar to everyone. It is similar to the user agreement we once read, whereby we were only allowed to install new software after we accepted its terms. Most software will be sold with such a statement, "VENDOR

DOES NOT WARRANT THAT THE FUNCTIONS CONTAINED IN THE SOFTWARE WILL MEET YOUR REQUIREMENTS." The same thing happens during Taobao's transactions, which is flooded with counterfeit goods.

Conversely, when all behaviors have a clear and standardized system that defines the boundaries of responsibility, perhaps things will become more harmonious. In the face of standardization under the requirements of laws and regulations, 4D printer manufacturers will seek to certify printers. At the same time, manufacturers of 4D printing raw materials will also try to meet or exceed the minimum material performance standards for manufacture and production. Meanwhile, 4D printing source designers will also constantly check and improve the accuracy of the digital models.

SECTION 9 | LOW ENERGY CONSUMPTION FOR THE EARTH

In light of the world energy and resource crisis, the destruction and waste caused by traditional manufacturing remain high. The advent of 3D and 4D printers at this time is being entrusted with a mission with critical and difficult timing.

Compared to traditional manufacturing methods, the materials used in 3D and 4D printing almost have no waste, which not only saves costs but also brings environmental benefits. It can also realize the localization of product production, reducing the need for transportation. Users only need to pay for the "design" and can print the needed products locally, thus reducing the demand for long-distance transportation and greenhouse gas emissions. Undeniably, product manufacturing based on 3D and 4D printing technologies will reduce material waste and carbon emissions in many aspects.

Indeed, the manufacturing of products based on 3D and 4D printing technology will reduce the waste of materials and carbon emissions in many ways.

The performance of 4D printing, in particular, is even more outstanding. It only uses the materials that make up the product and does not need any extra materials, but it also maximizes the utilization of raw materials in the manufacturing process. As demonstrated by Stratasys staff, it is possible to send the three-dimensional image data through a computer to a 3D printer to directly "print" the three-dimensional image into a physical object, as per the function of 3D printers. It works in much the same way as an ordinary printer that is connected to a computer, with liquid or powder printing materials, which then "printing materials" layer by layer through computer control to eventually turn the blueprint of the computer into a physical object. The entire production process is a bit like building blocks; materials are added layer by layer until the final shape is complete.

According to Wang Xianggen, General Manager of Stratasys-China, the 3D printers they currently produce are mainly for R&D and rapid manufacturing purposes. "Because 3D printing is suitable for various small batches, and personal-customized products, for a long time, 3D printers have been mainly used in R&D. However, with the maturing and cost-reducing technology, more and more fields are starting to use 3D printing to produce end-point products." Today, Stratasys has been providing 3D printing solutions for a wide range of industries to optimize traditional manufacturing processes. The range of customers covers aerospace, automobile, medical, and dental to education spheres, consumer goods, consumer electronics, and other fields. Wang Xianggen predicted that as the number of materials available for 3D printers increased, there would be more fields that 3D printing technology could be directly applied to for rapid manufacturing.

Joshua Pierce, a Michigan professor, has published a report stating that the direct application of 3D printing in the manufacturing field is not only cheaper but also more environmentally friendly. He found that the energy used by an object produced by a basic 3D printer was 41% to 64% less than the one produced by traditional technology. It also saved on the use of raw materials. For example, some molds produced by traditional technology were usually solid, while 3D printing could produce partially or even completely hollow molds.

More importantly, the localized manufacturing of 4D printing will over-all reduce the amount of products' transportation. Since the beginning of the Industrial Revolution, people have been increasingly relying on complex production technologies, and the places of manufacture were far from where people lived. Since human beings became obsessed with globalization in the 1980s, the places where goods were produced and sold could be tens or even hundreds of miles away from each other. Now that people can buy goods thousands of miles away, this has become so common that a significant portion of a product's value goes to transportation. Perhaps, economists and corporate accountants who are more concerned about economic benefits do not add the petroleum supply reduction and climate change to their theories or accounts. Yet, it is undeniable that these issues have become a heavy burden to social development and even the global environment. With the popularity of 3D and 4D printing, people will be able to manufacture many things locally and sometimes even directly produce and assemble products at home. As a result, a wide range of local businesses supported by 3D and 4D printing equipment will be more capable of meeting the needs of local customers. This means that the current demand for importing a large number of goods from overseas will also gradually disappear. The only thing left that will need to be transported may be raw materials, which take up less space during transportation.

SECTION 10 | PRODUCT REPAIR AND MATERIAL RECYCLING

The use of 4D printing will increase the service life of the designed and manufactured products while reducing the loss of raw materials in the manufacturing process. The products' storage of digital objects and localized manufacturing would not only minimize the problems in the product design process but also highly improve the convenience and possibility of product repair.

The concept of 4D printing, based on the 3D printing equipment and the storage of digital objects, will correct and improve the problems in product design. The iPhone charging dock, called an Elevation Dock, is a typical case. Elevation Dock is a successful innovation that raised funds through the Kickstarter website. At the time, an Indian designer designed a perfect iPhone charging dock, which successfully raised US$1.4 million, and then he delivered the charging dock to the sponsors. But it was the late delivery that created a very serious problem, which was that when the iPhone 5 was officially launched, it was using the new Lightning connector.

A London software engineer, Mike Hellers, explained his miserable situation, "I actually received mine the day before I received the iPhone 5. Couldn't have been worse, really, as it meant that I used the dock with the 30-pin connector for exactly one day before it became a useless paperweight." However, as an enthusiast of 3D printers, Mike Hellers was not as pessimistic as other sponsors. Instead, he took this as a challenge. He opened Tinkercad, a web-based 3D model design software on his computer, turned on his MakerBot 3D printer, and then began to design the adapter. It didn't take him long to post his designed adapter on the website Thingiverse. He then interacted with people who didn't have a 3D printer on Shapeway, and more than 12,521 users downloaded the adapter and were able to continue using the existing charging dock.

Matt Haughey, the founder of Metafilter, saw the adapter design on Thingiverse and thought there might be room for improvement. So, he and his friend Michael Buffington printed out some samples and found that the design couldn't hold the wires flat, which made the adapter easily fall out. Their improvement managed to solve the problem. They printed out a sample of the new design for testing and then shared it on Thingiverse. Haughey said, "It only took a couple of hours of tinkering through two existing designs to meld the best parts of each into something that worked better than previous designs."

A user of a 3D printer came up with a solution within a few hours, but it took the Elevation lab more than one month to work out an official solution. Actually, in hardware manufacturing, five weeks to turn things around is usually a pretty good result. But compared to the users of 3D printers, waiting five weeks is too long. "Any industry that involves physical objects and small parts is ripe for 3D printing." Haughey firmly believed that high-definition 3D printers would open up a whole new industry and play a critical role in repairing the components of used products.

In the past half-century, people in developed countries have been constantly forced to replace their commodities. It may take quite a long time for the older generation to save enough money to buy daily necessities. However, most people from the younger generations could choose to buy commodities via loans, and if the commodities do not work before they are paid off, that would be unacceptable for them from both, a personal or environmental point of view. Besides, the discontinuation of the components' production repairs the "can" to the "impossible."

Furthermore, the current market is full of products with similar functions mainly because only a few products contain replaceable components. We can blame poor product design, and also the fact that most manufacturers are expecting us to keep buying new products rather than repairing them for us. Therefore, even if a product has excellent repair service, it is not cost-effective for the manufacturer to just replace the components.

If the local 3D and 4D printing equipment are widely distributed, they will change the status quo of product repair. Most products' components could be stored digitally and then manufactured as needed. Even if some components cannot be found on the Internet, a smart person (or a future repair shop) will be able to design the component or scan the damaged part and repair it digitally on the computer, and then print out the component accordingly. As Jay Leno recently pointed out at Popular Mechanics, when he wanted to buy a component for a 100-year-old car, he could simply scan the component and have it produced through 3D printing or 4D printing.

All these behaviors, which are "superpowers" of the traditional manu-facturing model, will gradually become many people's instinctual reactions in the current era of 3D printing and the coming era of 4D printing. It is also because of this that when our watch stops working, our mobile phone is broken, or our laptop is too old to run, we could print a new component for replacement to make sure we do not "abandon" it due to the lack of components.

Not only that, but when we don't want some items anymore, we can recycle them through 4D printing, restoring their raw materials for rebuilding to maximize the use efficiency of raw materials as much as possible.

A young inventor, Tyler McNaney, created a "Filabot," which is a "plastic recycling machine" that grinds various plastics, such as milk jugs, soda bottles, all kinds of discarded plastics, and old 3D printed objects, then makes them into new usable printing filament materials. This will allow 3D and 4D printing to possess really strong sustainability qualities.

9

THE FUTURE (PART 1)

The Next Wave: 4D Printing—Programming the Material World, a research report by the Atlantic Council of the US, discussed a new disruptive technology, 4D printing. This was the first time that the scientific community formally discussed and deliberated the issues of 4D technology and defined 4D printing technology as the next wave, representing human society moving toward a bright future. There is no doubt that 4D printing will transform the traditional manufacturing industry and reshape future commerce with its advantages such as high-energy efficiency, low cost, and personalization. In the development of 4D printing technology, the scientific community will take the lead in solving constraints like equipment, materials, software development, and other aspects, to promote the popularization and application of 4D printing technology.

SECTION 1 | THE TEN ADVANTAGES
OF 4D PRINTING

The transformation that will be caused by the emergence of 4D printing technology will be far greater than that of 3D printing technology. Its impact on business forms and the transformation it will bring to the manufacturing industry will be far-reaching. For the manufacturing industry, 4D printing technology will bring at least the following ten advantages:

- SIGNIFICANTLY REDUCED MANUFACTURING COST

With the current modern manufacturing technologies, whether "Industry 4.0" in Germany, "Industrial Internet" in the US, or "Made in China 2025" in China, the cost of either manufacture or assembly of complex and personally customized products is relatively high. The cost also rises in proportion to the complexity of the components used.

However, when it comes to product manufacturing supported by 4D printing technology, the complexity of the components, and the structure of the product becomes less important. Because of the integrated printing of different components of the whole product, together with the ability of self-assembly, the assembly cost will be minimized, and the product manufacturing cost will be significantly reduced.

- UNCHANGED COST OF PERSONALIZED CUSTOMIZATION

Under the present manufacturing technology and environment, the cost of small-batch customization is still high. Even if we fully implement the technologies of Germany's "Industry 4.0" or China's "Made in China 2025" tech today, we still cannot solve the problem of the high cost of small-batch production.

But with the intervention of 4D printing technology, the cost of personal customization will be on par with that of traditional mass manufacturing, and it may be even cheaper. This is because the considered complex structures in traditional manufacturing will be simplified with 4D printing. Particularly, the cost of 4D printing will not fluctuate according to the complexity of components or the labor cost of assembly.

THE FUTURE (PART 1)

- THE RELEASED MANUAL ASSEMBLY COST

A key difference between 4D printed products and 3D printed products is that some 3D printed components need secondary assembly to be completed. 4D printed products do not need to be assembled by manufacturers or users. Manufacturers only need to print out the desired designed product by the user and then deliver the printed product to the designated location.

The user can directly trigger the medium according to their demand to enable the product to self-assemble, which therefore replaces the method that currently relies on manual assembly and manual disassembly. The labor cost will therefore be greatly reduced by the change of production methods brought about by 4D printing technology.

- ZERO-INVENTORY PRODUCTION METHODS

For manufacturing enterprises, all inventories are assets that will fluctuate with the changing market. Usually, except for special products such as alcohol, the longer the products are kept, the lower the value of the products becomes. For today's business forms, inventories are also a core part of an enterprise's capital cost. Once sales drop, it will directly lead to a decrease in capital turnover, thereby affecting the enterprise's profits.

Production through 4D printing will effectively alleviate this problem. Manufacturers could provide product designs based on users' ideas at any time and then print out the corresponding products. In this way, the traditional method of inventory sales will be replaced by the method that can achieve immediate buy-as-made and make-as-sold models.

- ENLARGED CREATIVE SPACE

To many designers, the most painful point is that creativity is beautiful, but the reality is cruel. A lot of creativity is constrained by various traditional processing and manufacturing techniques. A design that looks like a work of art is often changed into an entirely different-looking product.

Therefore, with no exaggeration, as long as designers can come up with creative ideas conforming to physical principles, 4D printing technology will be able to turn them into reality without compromise. This will allow the value of the creative design to fully bloom. So, for 4D printing technology, different creative ideas at the technical level will have the same degree of difficulty.

- LOWERED MANUFACTURING EXPERTISE

Today's simple or complex manufacturing has put forward requirements of expertise and proficiency to manufacturers to a certain extent, especially in some manufacturing fields requiring high expertise. Besides, it has also placed considerable demands on the workers' expertise and proficiency.

Why do many products manufactured in China, particularly high-end equipment, always lag behind some European and US products? It is because the skill level of our manufacturers is behind theirs, and the cultivation of technical personnel requires years of training and accumulation. It not only takes a lot of money to cultivate technical personnel but also bears the risk of brain drain. However, the application of 4D printing technology will reduce the specialized skills required for manufacturing complex components and will help us take on the manufacture of highly sophisticated crafts. Thus, the expertise threshold of manufacturing and the risk of brain drain will be effectively lowered.

- SIMPLIFIED MANUFACTURING PROCESS

Products nowadays are often assembled using multiple components, and different components require different equipment to manufacture them. For manufacturers, this not only needs a large space but also requires large capital investment in building equipment.

However, 4D printing technology is different. Just with a printer, based on different materials and the different shapes of the components set by the user, it can directly print out the components or the whole product. In particular for the whole product, as long as the various assembled and transformed models are embedded in the initial printing model, the printed product can self-alter to completion, even with the most complex assembly method.

- DEFECTIVE RATES ARE A THING OF THE PAST

The control of products' defective rates by traditional manufacturing enter-prises is a vital indicator of the enterprises' economic benefits. In the era of

4D printing, the term defective rate may disappear as part of the history of the manufacturing process, and the key to determining whether the products meet the requirements of users will shift to the process of design.

From a theoretical perspective, as long as there are no problems with the printer, printing materials, and the designed model, the printed products won't have any defects. In the future, the critical factor to determine whether a product is of good quality will no longer rely on manufacturing, but the design of designers. Therefore, the bad or good quality of 4D printed products will result through design, rather than manufacture.

- INFINITE COMBINATIONS OF MATERIALS

Combining different raw materials into a single product is difficult for today's manufacturing techniques because traditional manufacturing machines cannot easily combine multiple raw materials during the cutting or mold-forming process. Even though the molding multi-material hybrid injection has been applied in some fields, its cost and defective rates are high. The advent of 4D printing technology, however, has made it possible to mix and print multiple materials on one device, no matter if the materials are plastic, metal, or other synthetic. This means that in the future, a complete and transformable car could be produced just by printing.

- PERFECT CONSISTENCY IN MASS PRODUCTION

Although modern manufacturing techniques guarantee the consistency of production to a certain extent with the help of molds, there are still a variety of differences between products. It is difficult to ensure 100% consistency and accuracy. However, this problem will be solved with 4D printing, where mass production is as simple as copying a digital file and as consistent as a digital copy. 4D printing technology will also be able to ensure the consistency of production in both large and small batches.

These advantages are not part of science fiction, and 3D printing technology has achieved some of the above functions. As 4D printing technology continues to mature, it will bring us into a new industrial age where our manufacturing methods and lifestyles will be rewritten by printing.

SECTION 2 | THREE MAJOR CONSTRAINTS TO 4D PRINTING

Judging by how 3D printing has developed so far, behind its seemingly bright prospects, it still feels that it is short in self-appreciation. The constraints in its device upgrades, such as printing speed and stability of printing effects, have affected how a 3D printing device can effectively play the role of an interconnected media platform in the IoT. Besides, it also fails 3D printing when favored by venture capital.

Similarly, 4D printing, which is inherited from 3D printing and has just embarked on its early stages of development, has been facing a series of objective bottlenecks, such as equipment, materials, and techniques. How can we overcome the constraints to transform 4D printing into a breakthrough? This is not only the thinking of cutting-edge and frontline technologists but also where the opportunities are for emerging technology entrepreneurs.

So far, the bottlenecks that constrain the development of 4D printing are mainly in the following three aspects.

- EQUIPMENT CONSTRAINTS: BREAKTHROUGH FOR THE POPULARITY OF 4D PRINTING TECHNOLOGY

In the world of 3D printing, we have to use large-scale machines to print out the large-scale materials used for large-scale projects. These include construction and pipelines, which place high requirements on the printers, including volume, accuracy, and reliability, among other factors. This has created a real-life barrier to prompt and popularize 3D printing technology.

In the face of the lessons learned from 3D printing, 4D printing has taken another path, which is relying on "smart" materials to give printed objects the ability to self-transform. This largely alleviates the constraints of printing devices to technological development. But "smart" printing also means higher technical requirements for printing devices, which will inevitably lead to their high prices. This will, in turn, result in 4D printing having to go through a long period of being a "plaything of the rich" before it moves toward technology development and popularity.

Other than the price of small-scale 4D printers, also known as household-level printers, printing accuracy, operational expertise, and other aspects are still in the stages of being fully developed. These factors that currently constrain the rapid development of 4D printing have also left an uncultivated land for entrepreneurs interested in 4D printing. How could we effectively manufacture equipment and enter the market precisely to determine the success probability of entrepreneurs?

- "SMART" MATERIALS: PIONEERS OF 4D PRINTING

In addition to the constraints of the printing device itself, 4D printing is facing a more severe problem than 3D printing—the requirement of special printing materials. Compared with 3D printing, the requirements for 4D printing are much higher.

First of all, what 4D printing requires is not ordinary materials, but smart materials with memory functions (Figure 9-1), which are a new type of functional material that can sense external stimuli and then self-transform and self-assemble through judgment.

Figure 9-1 Memory smart material

Second, the materials need not only to be 3D printable but also have a sensing function, feedback function, information recognition, accumulation function, response function, self-transformation capability, self-assembly capability, self-diagnosis capability, self-repair ability, and super adaptability, as well as capabilities of rapid-response transformation and assembly.

It is foreseeable that in the future, with the advancement of science and technology, we will be able to realize the materials' self-transformation by triggering the stimuli of light, sound, heat, water, gas, temperature, and any other arbitrary medium. That is to say, while 4D printing faces infinite challenges on the long road to success, it also faces endless possibilities and opportunities. Just as there is a risk of failure and a probability of success for each particular experimented material and test, the selection of 4D printing materials and the attention and investment to their applications will become an important direction and development path for entrepreneurs.

- THE R&D OF SOFTWARE: ADVANCING 4D PRINTING

The reason why 4D printed objects can self-transform and self-assemble at a specific time through the triggering of the medium, other than the printing material itself and printing technology, is the design of the software used to print them. This means, while completing the design for the printing materials, the transformable elements will be directly implanted into the material. A simple but not very accurate understanding is that 3D printing means designing products, while 4D printing is embedding the printer in the printing model. Therefore, for 4D printing, design software is critical and is one of the links in the development that needs to be advanced. How can we de-professionalize design software so that users can operate and apply it without prior knowledge? This is one of the indispensable factors for the popularity of 4D printing.

With the gathering of capital and talent, much like how they have once advanced 3D printing, the problems that constrain 4D printing will soon be solved, too. This will surely lead to a new round of opportunities for the manufacturing of 4D printing devices and the development of application software. More importantly, it will promote and drive the development of new materials.

SECTION 3 | SIX MAJOR TRENDS IN 4D PRINTING

On top of 3D printing, a self-changeable dimension is added to it for 4D printing. This brought profound changes and influence on the development of both industries and human society. When such science fiction technology enters our lives and becomes a reality, it will affect and change the current business forms and will trigger new development trends.

4D PRINTING WILL TRANSFORM INDUSTRIES

Whether it is Industry 4.0 or 3D printing, what they have solved is only the shift from the past mass production to customization, but not the products' self-assembly and self-transformation needs. Although 3D printing technology has been a more direct help for the design of complex structures, including making some structural components lighter and stronger, it is still unable to self-drive and change in response to external environments.

Despite the fact that 3D printing technology has been applied in industries such as aerospace and construction, it is still difficult for it to meet the changes in different environments. For example, a 4D printed aircraft can decompose itself while facing specific environmental changes to provide the best protection to its passengers. The same could be done in aerospace, whereby we wouldn't have to carry the huge assembled equipment into space, but merely the materials and a 4D printer, or the printed materials, and under the trigger of a specific medium, the space station, space suit or other operating equipment could automatically assemble and form for us.

For everyday industries, 4D printing technology will revolutionize the current manufacturing forms and business forms.

4D PRINTING COULD HELP TO OVERCOME CANCER

The feasibility of using 3D printing to print organs and limbs and effectively match them to patients has been proven in the medical field. The same is also possible for some relatively bony limbs. However, when it comes to the printing of some soft tissues and organs, the performance of 3D printing technology is not as amazing as that of 4D printing technology.

The most disruptive aspect of 4D printing technology is not the printing of organs or limbs, but the printing of cells. By printing drugs in the form of cells that are nanoscale and similar to human cells, and then injecting them into the human body, and setting the trigger media as the corresponding cancer cell virus, 4D printed drug cells could patrol the human body. They will be able to release the drugs contained point-to-point to kill the encountered cancer cells. The hope of humankind to overcome cancer perhaps is pinned on 4D printing technology.

For unknown viruses that humans are exposed to, with the help of 4D printing technology, the same method can be adopted, which will inject a harmless vaccine into the human body.

The Popularity of Private Factories

Before 3D printing technology, personal customization was almost impossible, and all our consumption could only depend on industrial mass production. For manufacturers, to reduce costs and improve enterprise benefits, they needed to integrate supply chains for different components and adopt large-scale industrialized, standardized, and automated production. Therefore, it was not easy to meet the demand for personalization.

But when it comes to 3D printing, it is easy to fulfill the demand for personalization or even personal customization based on users' preferences. However, what 4D printing can change is not limited to personal customization. More importantly, it could allow each user to own a "factory" that can define and produce the items they want. Imagine you want to buy a refrigerator. You don't have to buy an entire huge refrigerator for your home, but you could buy the core components, or you can even define and print the components out on a 4D printer, and then the printed components will automatically assemble into a refrigerator for you.

This is very different from personal customization, whereby we can express our ideas, build our lives, and even define our innovative "factory" at home with the help of 4D printing technology. What's even more amazing is that when we need more space for the freezer compartment and less space for the cooler compartment, the refrigerator could automatically trigger to assemble and adjust the space of the compartments accordingly.

The Speed of Innovation Will Be Further Boosted

One problem faced by traditional industrial manufacturing is the long time cycle from product establishment to development and then to market launch, which usually takes at least 6 to 24 months. This is because it involves the stability of batch manufacturing, as well as the manufacturing and service life of molds. All these issues constrain the time cycle of product innovation.

But with the emergence of 4D printing technology, not only will our ideas be printed quickly, but we can also embed a variety of possible corrections into the models. After the object is printed out, we can trigger it to self-transform, self-refine, and self-correct according to our needs, which will indeed be a meaningful thing for people who are pursuing innovation, personalization, and differentiation.

The rapid realization of personal ideas will further reduce the time cycle of innovation, and much the same will be done for traditional industrialized enterprises. Perhaps shortly, what traditional factories sell will not be the products, but rather years of technologies and services. Users could purchase technical and manufacturing experiences to assist them in the innovation of products based on their ideas.

The Appearance of New Business Models

With the arrival, maturity, and popularization of 4D printing technology, we may not sell the completed products in future supermarkets but sell some key components instead. So, users could buy the needed components and then connect them with the printed ones to realize innovation.

The business models brought about will cause significant changes. On the one hand, the current sales model of supermarkets will change from selling completed products to providing 4D printing customization services and technical support. On the other hand, people engaged in creative work will obtain commercial value by providing consumers with creative designs and 4D printing services. Besides, small chain supermarkets mainly selling a variety of components will emerge.

More than bringing new business models, IKEA, which focuses on selling creative designs and leaves the assembling job to customers, 4D printing will completely change its business model. 4D printed products will not need any manual assembly.

The Issue of Intellectual Property Will Be Magnified

4D printing and 3D printing are both facing one common controversy, that is, copying copyrighted product designs. As such, this issue of intellectual property will be magnified. When a user sees a creative work they like, they could take a picture of it, generate the corresponding 3D model by using software on the computer, and then input the model into a 4D printer for printing. This is when the protection of the intellectual property will become extremely tough.

In particular, 4D printed products could even self-assemble and self-change, which increases the difficulty of obtaining evidence. Intellectual property protection is a new issue faced by the new technology era.

10

THE FUTURE (PART 2)

The maturity and popularization of 4D printing will bring not only innovations to human manufacturing technology but also transformations to the future social ecology. The process of making dreams come true will comprehensively stimulate the innovations and creative ability of all people, including the field of children's education and development. It will be a brain-opening spectacle.

With 4D printing, there will be nothing you can't do, only what you can't imagine. The future world will be transformed beyond human imagination.

SECTION 1 | WORLD SHAPED BY 4D PRINTING

In a world shaped by 4D printing technology, our tomorrows will be distinct from today. These changes will permeate every moment of our daily lives and infiltrate every sector of our industrial, agricultural, and commercial production.

As the first rays of morning sunlight touch the bedroom's floor-to-ceiling windows, musical curtains will gently rise. Accompanied by soft melodies, cells within our bodies stretch lazily—I am awake.

Yawning, the neatly arranged teeth shed accumulated debris upon contact with air, instantly freshening the breath. Stepping out of bed, the flowing, sensual nightwear transforms into a beautiful outfit triggered by heartbeat and pulse, bringing instant comfort.

Entering the living room, a smart robot has already prepared a 4D printed exquisite breakfast—a tailor-made, additive-free natural health food that reserves a day's energy within our bodies, controlled and released by a personal health guardian.

Energized, we board a floating mini spaceship, transformed from a smart robot, instantly connecting our brainwaves to the team's mind map workspace. Here, intellectual collisions spark inspiration and miracles. As brain map data is generated during these intellectual exchanges, creative models are simultaneously transformed into physical objects by the 4D printer, evolving with our changing thoughts without any repetition or waste.

Though thinking can be exhausting, our internal health guardian ensures we never overwork. Before fatigue reaches a certain threshold, it activates a "snooze bug," clearly signaling an impending need to rest and rejuvenate.

After a day's work, entertainment is essential. In malls, modern 4D printing equipment is everywhere in our lives, with personalized small 4D printers for sale, making production and manufacturing a spontaneous act of "personal customization."

As night falls, glamorous workwear transitions into casual home attire. Lying lazily on the sofa, wearing a mind map headset, we archive and organize thoughts for real-time retrieval while also importing new knowledge and information, continuously enriching ourselves.

In the world of 4D printing, our tomorrow will never be another today. Every day will be unique, constantly gathering and iterating upon the wisdom of humanity for self-improvement and upgrading.

Simultaneously, our industries, from manufacturing to medical diagnostics, will also become distinctive through the development of 4D printing technology.

SECTION 2 | HUMANS' CREATIVE ABILITY STIMULATED BY 4D PRINTING

3D printing makes it possible for Makers, designers, and creative people to express and actualize their ideas. It also makes these ideas commercially valuable. It is a consensus before 4D printing technology that to realize the dream of becoming a Maker may have to adopt the development of 3D printing technology.

Experts believe that desktop 3D printers will have an impact on the real manufacturing industry or its industry chains, especially for household goods. This point of view is based on the convenience of 3D printing and the technical principles that can rapidly actualize users' ideas. However, 4D printing technology will not only disrupt the manufacturing industry or its industry chains but also stimulate humans' potential creative ability on a much deeper level.

The four dimensions of 4D printing are beyond our recognition of three dimensions, so 4D printing itself is a kind of creative technology. Its critical difference from 3D printing technology is that under the action of an external triggering medium, 4D printing will self-transform and self-assemble, which will effectively stimulate humans' creation ability and bring along the following changes.

Releasing Human Labor

This can be understood on two levels. The first is that the inspiration of most designers and Makers is currently actualized by handcraft. The problem brought about by this method is the limitation of handcraft itself, which means the handcraft masters have to experience years of honing and settling into their craft, and even one technique may require years of work. Even if the handcrafted technique has been mastered to a very advanced level, it is still almost impossible to produce multiple identical products, and this is the limitation of the designer's creations.

Some people may think that this kind of scarcity will make the product precious. But, for those masters who integrate aesthetic creations with commercial value, certain basic handcraft techniques take too much time, which is a great limitation to their creation value. 4D printing technology not only gives designers more creative space but also a choice to let the handcraft masters further build up the 4D printed models with their excellent techniques.

The other level is the difficulty of the second replication. With the integration of 4D printing technology and 3D scanning technology, the second replication can be realized by fusing it with 4D printing technology in the early stage of creation. Alternatively, for handcrafted products, 4D printing technology can be used to produce the second copy.

Bringing Dreams into Reality

Leaving aside 4D printing technology, it is difficult for individuals to express their ideas in the physical world and make those ideas become usable commodities without the use of 3D printing technology. Perhaps we can use traditional machining techniques such as turning, milling, planning and grinding, or machining centers to process the three-dimensional physical objects, but this is still difficult for some complex components.

4D printing technology fundamentally solves the problems of forming arbitrary structures and shapes, allowing self-assembly capabilities of some products to assemble and save on labor. More importantly, the problems of transportation and storage are solved as well. Take a table as an example; its transportation usually happens in two ways using traditional methods. One is to disassemble the table into small unit parts, transport them to the designated location, and then manually assemble the unit parts into a table. The other is for the seller to assemble the table first, and then transport the whole table to the designated location. The disadvantage of the second way is that the table occupies a huge space in the transportation process. However, this disadvantage doesn't exist in 4D printing technology because we can print the table layer by layer into a plank first. Then, after transporting the plank to the designated location, it will self-assemble and self-transform into a table by triggering its medium.

With 4D printing technology, the Transformers, the magical paintbrush of Ma Liang, and the Monkey King's *Jingu Bang* in our minds will no longer be stories. Designers, Makers, or the general public, as long as we have creative ideas, 4D printing will make them a reality.

Furthermore, the maturity and popularization of 4D printing technology will further stimulate the innovation and creativity abilities of all people, including their application in the field of children's education and development. It will bring us a brain-opening spectacle. Perhaps in the future, we only need one book, which will become a corresponding textbook in different classes. Human creativity will be influenced by the application of 4D printing technology, and human potential will be continuously stimulated as well.

SECTION 3 | LEADING THE NEW ERA OF "SMART MANUFACTURING"

The advent of 3D printing, a precursor to 4D printing, brought about disruptive changes to industrial production and economic organization models, deeply impressing upon us that the fourth Industrial Revolution is on the horizon.

Imagine tomorrow, the disruptive creativity of 4D printing technology is a form of intelligent, adaptive innovation that will completely alter traditional manufacturing methods. With the initial formation of the 4D printing industry chain and the preliminary large-scale application of this series of technologies in public infrastructure, the fourth Industrial Revolution will officially commence.

As a brand-new Industrial Revolution, the fourth Industrial Revolution builds upon the previous three, taking new steps forward by leveraging "smart manufacturing" technologies, including 4D printing, to significantly enhance resource productivity. From a historical development perspective, considering industrialization, it's clear that China is catching up with the dawn and mobilization phase of the fourth Industrial Revolution.

THE FIRST INDUSTRIAL REVOLUTION: THE PRIMITIVE ACCUMULATION OF LABOR AND TECHNOLOGY

An adequate labor force, rich technical experience, sufficient funding sources, and a vast market space together formed the standard configuration of the first Industrial Revolution. At the same time, a world-famous, vigorous "enclosure movement" fueled the first Industrial Revolution that erupted in England.

When a large number of peasants lost their land, income sources, and basis of survival, they were forced to flock to cities in search of the last bit of hope for life, becoming an important source of labor for new urban industrial fields. Struggling for survival on the line of life and death, the lower labor force, British society's lower and middle class merchants, craftsmen, and workshop owners, all possessed a Puritan-like hard work and entrepreneurial spirit. Many of them also had another identity—technicians with rich technical experience, making the technological development of the first Industrial Revolution a possibility. According to historical data, most inventions in the textile, mining, metallurgy, and transportation industries involved in the first Industrial Revolution were not made by scientists but by technicians.

At the same time, British capitalists had already accumulated a large amount of capital through methods such as "piracy, commercial wars, colonial plunder, and the slave trade" before the Industrial Revolution. This made capitalists willing and able to financially support and invest in platforms that could improve production efficiency and profit levels. Moreover, Britain was the most powerful colonial country in the world at that time, with continuously expanding over-seas markets. The huge market demand could not be met by manual production alone, and capitalists driven by profit would inevitably seek all possible ways to improve efficiency and increase production. Thus, the Industrial Revolution, represented by the steam engine, was about to emerge—Watt's improved steam engine thereby ushered human society into the Steam Age.

THE SECOND INDUSTRIAL REVOLUTION: "SCIENCE" OFFICIALLY REPLACES "EXPERIENCE"

After 1870, the development of science and technology advanced rapidly, with a plethora of new technologies and inventions emerging and being swiftly applied to industrial production, greatly promoting economic development and leading to the second Industrial Revolution.

The US, the creator of the second Industrial Revolution, achieved world-leading results through it. This success was related to its conditions, such as a vast treasury of raw materials, ample capital supply from Europe, a continuous influx of cheap immigrant labor, a huge domestic market, a rapidly growing population, and continuously improving living standards. Moreover, the second Industrial Revolution advocated two important methods: standardized manufacturing and assembly line production, which developed in the US. It involved manufacturing standards and interchangeable parts to be quickly assembled into complete unit products with minimal manual labor.

The biggest difference between the second and the first Industrial Revo-lution was not electrical machinery but the replacement of "experience" with "science." "Scientific manufacturing" could meet higher-level human needs for cognition and social interaction, and humanity began to value the development of science, thereby accelerating scientific progress.

THE THIRD INDUSTRIAL REVOLUTION: COMPUTERS + INTERNET

The rise of the third Industrial Revolution was marked by the application of nuclear energy technology, space technology, electronic computers, and the development of new high technologies such as synthetic materials, molecular biology, and genetic engineering. The US, with technological, material, institutional, and cultural advantages, became the leader of the third Industrial Revolution.

Technological conditions. In the realm of thinking technology, American pragmatism began to form; experimental technology was characterized by military-civilian and science-engineering integrations in production, electrical and aviation technologies were leading.

Material conditions. The US had superior natural resources and a huge domestic market favorable for mass production; it was untouched by both World Wars on its soil, avoiding any war damage while earning vast amounts of money from arms trades.

Institutional conditions. As the first bourgeois-democratic constitutional country, the US allowed scientists' autonomous creativity with minimal political intervention and high freedom for scientists, leading to the birth of more versatile inventions.

Cultural conditions. Americans, coming from all over the world, integrated various cultural traditions; during World War II, it also attracted a group of excellent European scientists, such as Einstein and von Neumann, driving the training of their own talents and establishing various academic organizations and a diversified scientific research system.

Following the US, China took solid steps in reform and opening up during the third Industrial Revolution, actively introducing and absorbing advanced scientific and technological management methods worldwide, rapidly catching up with the world's scientific and technological revolution, and significantly promoting the country's modernization process.

The most significant innovations of the third Industrial Revolution were the computer and the Internet. Previously, whether machinery or electricity, operations always required human control, limited by space and time. However, the invention of computers and the Internet made automation and intelligent manufacturing possible, finally enabling the manufacturing industry to break free from spatial and temporal constraints.

The Fourth Industrial Revolution: 4D Leads
Intelligent Manufacturing

Relying on the computer and Internet-represented "network manufacturing," the real significance of 3D printing, "personalized manufacturing," and 4D printing, "intelligent manufacturing," began, making it possible.

In the current world where the population and middle class are rapidly expanding, one of the long-term potentials of 4D printing is to provide richer products and services with fewer resources, thereby promoting a truly sustainable world. This has led professionals to call 4D printing technology the core of the "Green Industrial Revolution," truly a "green additive manufacturing" society.

As is well known, traditional manufacturing often involves "subtraction," where materials are shaped into finished products through cutting, trimming, and polishing, inevitably wasting raw materials and causing pollution. However, manufacturing with 3D and 4D printing technologies is an "addition": stacking powdered and other fine materials layer by layer to form the final product. Moreover, with the support of 4D printing technology, the self-transforming function of objects contributes significantly to meeting future societal needs.

Imagine if objects could transform or change their properties according to personal commands or preset programs; the world would benefit immensely. For example, airplane wings that automatically deform according to airflow changes, furniture, or even buildings that can self-assemble and disassemble for different functions. If so, earth's limited resources could be better preserved, and objects could be more fully recycled: through commands, objects could be decomposed into programmable particles or components, then reassembled into new objects or with new functions.

SECTION 4 | NEW CAREER PATHS FOR
INDUSTRIAL DESIGNERS

Design plays a key role in a country's economic transformation and upgrade, and industrial design plays a significant role in promoting the current upgrade of China's manufacturing industry. With the global economic slowdown in recent years and the rising cost of China's domestic manufacturing, many enterprises relying on low-cost competition have found themselves in a dilemma. How could they increase the added value of products through design? (Figure 10-1) The creation of competitiveness in Chinese manufacturing seems to be a concern for most enterprises in China.

Figure 10-1 Creative design

Moreover, impacted by the Internet, the traditional manufacturing industry, and the consumers' consumption philosophy have undergone profound changes. Concepts of personalized consumption and personal customization are on the rise. The most obvious of the changes caused by these concepts is the formation and emergence of groups represented by the Makers. They undoubtedly impact the traditional manufacturing industry, which is concentrated on the advantages of standardization and mass manufacturing. The continuous maturity of 3D printing technology provides the foundation for the realization of such creativity-led personally customizable consumption products (Figure 10-2).

For practitioners engaged in creative design, especially industrial designers focusing on manufacturing products, 3D printing technology is now used to produce and manufacture the ideas in their minds into physical products. This has become a new opportunity and a trend to materialize ideas and then realize their commercial value via e-commerce platforms.

Figure 10-2 3D printed cups

But in the era of 4D printing, everything will be subverted. Designers will become ubiquitous "make-up artists" and "beauticians" of products. This means that designers will soon be able to open up design studios with their style and characteristics on the cloud service platform, and any consumers could be their customers. Consumer could express their creative ideas via software tools on the computer, or they could choose a designer who fits their creative style and convey the ideas to the designer through text, pictures, voice messages, or any other form. Then, the designer will come out with the design in response to the consumer's request.

Alternatively, a consumer could post their ideas indicating the remuneration they are willing to pay on the platforms that provide creative design services, and any interested designer could accept the job and provide the corresponding service. For Makers after the creativity based on the technology is realized, Makers will dress up the products, allowing them to walk into consumers' everyday lives from the lab. The biggest difference between 4D printing and 3D printing is that there will be more possibilities and challenges to designers, such as the beauty of the product under different forms, or the best way to change products into different shapes. All these challenges will test the designers' wisdom and will also bring some changes to the design discipline.

SECTION 5 | THE POPULARITY OF XIAOMI

In China, the once-popular Xiaomi smartphone known by many people has become a classic case many visiting professors discuss. "Xiaomi gathers a group of reliable people who use the Internet to make smartphones." According to Lei Jun, the founder of Xiaomi, 600,000 enthusiasts who call themselves "Mi Fans" were involved in the development of the smartphone system. From this, the Xiaomi smartphone was attached to a large number of enthusiasts, and more and more innovations that meet the habits of Chinese people's use have been created by "Mi Fans" and put into use by the Xiaomi smartphone. The open design of the Xiaomi smartphone has firmly attracted 700,000 enthusiasts before its launch.

The same successful case of open design has also been proved by Nokia's 3D printing community project. On January 18, 2013, Nokia's 3D printing community released a "3D printing development kit" to help people design and make their own Lumia 820 phone case. As introduced by John Kneeland, a Nokia Community, and Developer Marketing Manager, the Lumia 820 smartphone has a detachable case, which has anti-shock and anti-dust protection. Customers can replace the case with different colors, and for the advanced version of the Lumia 920 and the intermediate version of 820, wireless charging capability could even be added to the case. These special features give most Lumia 820 users great options. Other than that, Nokia also released 3D templates, instruction manuals, recommended materials, and the best practices, so that anyone familiar with 3D printing could customize a Lumia 820 phone case for themselves. Only a few days after the announcement was posted to the Nokia community, a Nokia 3D printing challenge for the Lumia 820 was held to encourage people to share design ideas for phone cases that could be 3D printed and used as replacements. Six days after the release of the 3D printing development kit, the 3D printed Nokia Lumia 820 phone cases, containing some functional buttons, were displayed online.

The same creativity will be fully exploited in 4D printing. With the frequent evolution of digital models, product design, and production plans based on the storage and transmission of digital models will be continuously improved and upgraded. This will be done by the second innovation wave from different printing enthusiasts, and the printed products will be evaluated and tested during the printing process. Afterward, the enthusiasts will be able to upload the improved design plans and digital models, which will not only allow others to benefit from the design improvements but also give designers and other printing enthusiasts a new direction of thought.

With the development of computers and the Internet, the increasingly barrier-free spread of cyberspace will give birth to "open design" in the 4D printing era to root, breed, and further spread its ideas. As more and more people use design software with AI, 3D, and 4D printing technology, and more people modify and implant their ideas to modify and innovate the designed digital models through the Internet—they will gradually turn the future product design process from exclusive to an open one.

SECTION 6 | DDM MAKES CUSTOMIZATION POPULAR

Mass production is the best way to reduce production costs in a traditional manufacturing environment because the development cost of abrasive tools could be diluted with the increasing production. At the same time, the cost could become very large if the developed abrasive tools are only used to produce a few products, which is why customized products are often expensive.

The application and promotion of DDM will undoubtedly bring customized products into the homes of the common people. Everyone will be able to design and implant personalized elements into their products, and the printer will only perform the printing job according to the data model received. It will be just how we print our files on a 2D printer; it does not matter if the content of the files is numbers, Chinese characters, or colorful pictures.

Those who have visited the Cubify.com website are probably familiar with the case of customizing the Apple mobile phone model, which can produce a 3D finished product with a raised back based on the digital photo provided by the customer. Similarly, ThatsMyFace.com invites visitors to upload photos of the front and side of the face and then generates a colorful 3D head model, which can be built onto a plastic doll or Lego model. Moreover, with the continuous advances and maturity of 4D printing technology, the printed head model will be able to show signs of growth or aging in the natural environment over time. Such customized and changeable products will enable people to constantly feel the rapid changes of the world, and it will no longer be hard for us to easily own a small superhero figurine prototyped after ourselves.

There are already too many similar cases. Personalized custom design and low-cost product manufacturing will terminate traditional mass production. Instead, it will replace it with the low-cost production of disposable products or small batches of components because special abrasive tools will no longer be required to start the production. For example, the producer of the 2012 James Bond film, *Skyfall*, needed three Aston Martin DB5 model cars that were one-third the size of the actual car, and they printed them out on a 3D printer model Voxeljet VX4000. These model cars were 3D printed into 18 parts and then assembled and spray-painted before being turned into very realistic and expensive-looking replicas. Likewise, a custom motorcycle manufacturer called Klock Werks Kustom Cycles manufactured a motorcycle in five days, used engineers that designed the dashboard, fork tube cover, headlight baffle, pedals, pedal bottoms, and wheel gasket covers by using a 3D model software called SolidWorks. Finally, they printed out the model using a 3D printer. The speed of the finished motorcycle even set the land speed record of the US Motorcycle Association. In the foreseeable future, the parts of this motorcycle will even be able to achieve self-transformation, self-extension, and seamless splice under the action of a certain medium with the application of 4D printing technology. This will further simplify the printing process and improve the printing efficiency.

Many successful cases have proven to us that 4D printing technology relying on 3D printers can produce various personalized products quickly and conveniently. Plus, the products could transform and upgrade within the time dimension. The low cost and techniques of the products are unachievable by traditional manufacturing methods. For either enterprises or individuals who want to produce one or a small batch of the latest trendy products, 4D printing will be an unparalleled choice.

SECTION 7 | THE "MAGIC" EFFECT OF 4D PRINTING

4D printing, which follows the pace of 3D printing, will lead us to new dreams and make science fiction a reality. While Captain Jean-Luc Picard was sitting in the preparation room of the Starship Enterprise and wanted to have a hot drink, he just needed to verbalize it in words, and the "replicator" on the starship could collect all the necessary atoms, including the atoms to make a cup, and then make him a cup of hot tea. The common use of the "replicator" by Captain Picard is similar to our use of the microwave today. The microwave oven uses wireless radio waves to activate atoms and generate heat, which was an exciting event in the 1950s. While the unexplained magical technology of the "replicator" collects and aggregates atoms to form food and drinks.

Star Trek: The Next Generation is a science fiction film. However, Picard's "replicator" is not completely impossible to turn into reality. When you add a little poetic imagination to the current industrial 3D printers, you will see the "replicator" in *Star Trek*. In a basin of stationary liquid resin, a laser begins to track the shape of the resin-like lightning to produce various shapes, and then as if there was magic that made different kinds of food, it suddenly appeared. With the advent and popularization of 4D printing, these foods and items could even change according to our expectations and needs under different time dimensions and actions of media.

Let's put aside the poetic imagination as we are still far away from the self-aggregating atoms, or at least a considerable distance from their effective use. However, through 3D printers and 4D printing technology, we can realize the personalized custom production we need. As long as we can present ideas on the computer and shape them with a machine just by simply pressing a button, the objects will appear like magic. Perhaps this is the saying, "Any sufficiently advanced technology can be called magic."

Lower Costs: No More Expensive Customization Fees

When 3D printing was first introduced, some scholars used the analogy of a platypus.

When the platypus was first spotted, explorers thought it was a hoax, thinking it was a joker who stuck a duck's beak, webbed feet, and a kangaroo's pouch on a furry animal.

Similarly, 3D printing is the "platypus" of the manufacturing industry, combining precise digital technology, factory repeatability, and the craftsmen's design freedom. Just like the machines in the factory, 3D printers are also automated. The digital design file will concisely receive the instructions for producing a specific product, and then guide each step of the 3D printer. This process can be saved or sent anywhere via email. Just as craftsmen can produce multiple products, 3D printers have multiple uses. A 3D printer can produce a wide variety of products without a large upfront investment.

Following in the footsteps of 3D printing, 4D printing has the same effect. We can 4D print 1,000 different products and 4D print 1,000 identical products, and the costs will be the same. The cost of customization virtually disappears because the development and manufacture of abrasive tools will no longer be needed.

In other words, the application and promotion of 4D printing have removed the shackles of any conditions for the manufacturing industry's personalized and customized production. Even though traditional mass production has a high-efficiency advantage that can increase enterprise profits and reduce retailing prices, traditional economies of scale harm product diversification and customization. Conversely, craftsmen can easily produce many diversified and customized products, but the output will be extremely restricted. With the arrival of the era of 4D printing and digital manufacturing, we will be able to choose between mass production and customization without having to pay expensive handcrafting costs.

Along with the recognition of 3D printing, people's acceptance of 4D printing has rapidly improved. For enterprises whose business model is to sell a small number of unique, customized, or changeable products with high marginal revenue, the development of 3D printing and the application of 4D printing will undoubtedly bring revolutionary leaps forward. This will be of epoch-making significance.

More Efficiency: The Manufacturing of the Product's Prototype Can Be Broken Down

As our world accelerates, enterprises are increasingly eager to shorten the time it takes to go from the design stage to product delivery. The production cycle is a key measure of enterprise efficiency, which means that the shorter the time between the design and completion of the final product, the better.

4D printers will minimize product delivery time by enabling designers and engineers to efficiently produce prototypes at low cost on-site. The prototype is the first draft of the product, which helps designers, engineers, marketing teams, and manufacturers conduct multiple checks to see how the finished product will look. Previously, we provided rapid prototyping services to automotive manufacturers through 3D printing. By 3D printing out the design concepts and presenting them to the project team, the automotive manufacturers saved time and even saved consumers' time.

In comparison, prototyping was a lengthy and costly process in the past, and it was risky for manufacturers to take shortcuts or to be overly confident in the fact that product design could be realized. Even if the designs were beautifully made, manufacturers could still find many other problems after they were produced because it was difficult to visualize the designs in CAD models.

For example, if you own a car, you need to repair and refuel it. But we cannot simply make a hole in the engine to refuel the car, and human hands wouldn't fit like that, either. Therefore, we can imagine that 3D printed prototypes will soon replace those handmade with foam or clay. Many enterprises have skipped foam and clay prototypes and directly adopted 3D printing prototypes. After 4D printing is implemented, everything will be further sublimated.

Through 4D printing, we can even perfect the design and further shorten the delivery time based on 3D printing. The 4D printed prototypes can change at a specific time and under the action of external media so that the printed samples can more effectively and accurately fit the product design and application requirements. Instead of trying to achieve perfection by printing the product again and again. In this way, the time cost will be shortened, and production efficiency will be enhanced.

BETTER QUALITY: MANUFACTURING THE
BEST-CUSTOMIZED COMPONENTS

The printing of customized terminal components is one of the fastest-grow-ing applications of 3D printing, and the customized component will be one of the leading products of 4D printing, too. So, the customized components will no longer be prototypes but real products. People who have a 3D printer at home could log on to the community forum to exchange ideas and design documents of standard replaceable components, such as door handles and shower curtain rings, which could also be rapidly communicated, exchanged, and applied in 4D printing. The 4D printed components include door handles, gears, antiques, or discontinued components that are handmade and extremely expensive.

As production of customized components is not profitable in the economies of scale, small-scale and technical 3D printing service providers and 4D printing operators will start looking for new business opportunities. Automobile and motorcycle manufacturers (and even Mars rover manufacturers) are using 3D and 4D printed custom components to produce concept cars or machines. After all, a multi-million dollar vehicle is too expensive for test drives.

4D printed terminal components are widely used in the medical industry and the dental industry because the products of these industries require closer and more accurate integration with the human body. The suitable braces and tooth crowns for patients used to be customized, but now, they are increasingly 3D printed. In the future, the transformation capability of 4D printing will make it an even better choice. Likewise, by scanning a patient's ear canal or residual limb, the scanned data could be used to print out corresponding hearing aids and prostheses to maximally fulfill the needs of the patient under the action of time dimension and media.

CHEAPER: NO WASTEFUL CONSUMPTION AND
LOW-COST PRODUCTION

4D printing will minimize consumers' passive, wasteful behaviors. This is because 4D printed products could transform under the action of a certain medium along with the change in time and space dimensions. This will satisfy consumers' needs to the greatest extent. To give a simple example, the growth of newborns is "changing with each passing day," and whether it is clothes or shoes, there will always be the "embarrassment" of them being too small after not wearing the items too long. But 4D printed children's clothes or shoes will have the magic of being able to "grow up" with the children.

The consumption of 4D printed products can reduce unnecessary waste for consumers, and the production of 4D printed products can also reduce the cost of the product development process for manufacturers. Some 4D printed prototypes are used to demonstrate design concepts, while others are used to test stages of products' life cycle to figure out how to achieve mass production of the product's components. The "test and adjust" ability of 4D printed components can not only maximize the variations of the products in different time dimensions and under different media conditions but also minimize the inherent defects of the complex-designed products. This has also been fully proven in 3D printing.

The prototype for testing and adjusting can be the non-assembled components used by engineers in a simulated production process. Earlier, when Microsoft announced the top-secret product concept (the hybrid tablet and laptop the Windows Surface), the world was shocked. The media wanted to know how Microsoft had developed this product under confidentiality. Usually, when a company first announces the launch of a new, cutting-edge technological product, mysterious photos will leak out of its manufacturing factory. How Microsoft has done this is that the product's prototype was 3D printed in the hardware department located on a university campus, ensuring that product development was being carried out secretly.

The other purpose of testing and adjusting was to make sure that the machines in the factory could produce the exact product according to the design concept. In the engineering product design course, students spend weeks choosing between the great design ideas and the realities of the factory. Although those thick textbooks detail which design concepts are and are not feasible on the production line, it is hard for common molding machines or cutting molding machines to manufacture hollow objects, interlocking parts, or products with complex internal structures. Not all production challenges can be avoided under the guidance of the textbooks.

If an enterprise finds out too late that the components of the new product do not fit together, the initial investment will be wasted. The 4D printed product's prototype for testing and adjusting can help designers of mobile phone products assemble tiny hardware, such as hearing aids, car steering wheels, razors, combs, and smartphones, which must be comfortable to the touch or fit snugly on the user. Even though design software and computer simulation are becoming more and more advanced, even the best designs may not always be produced exactly as planned.

SECTION 8 | THE FUTURE WORLD IS BEYOND IMAGINATION

The most direct understanding of 4D printing is that the time dimension is added based on 3D printing, and then time and corresponding models to be transformed into the printing model are embedded with the help of software. This allows the transformable printing materials to automatically transform and assemble into the predetermined shape under the action of a certain medium at a specific time.

The biggest change is this process does not need to resort to complex electromechanical principles, or complex computer programs and electro-mechanical equipment. Instead, it prints a kind of self-transformable material, which is embedded with the assembling and transformable design capabilities through software. Then, at a specific point in time, the printed product can fold, assemble, and transform automatically into a shape conformed to the intention of the user's initial design. From this perspective, we can perhaps make a distinction between 3D printing and 4D printing, and the distinction is that the logic of 3D printing is to build the printing model and then print out the final product. While the logic of 4D printing is to embed product designs and time into the transformable smart material through a 3D printer. So, it can be said that 3D printing is a static printing process, whereas 4D printing is a dynamic shaping process. The key to 4D printing is its ability to self-transform over time.

The key to 4D printing is not the innovation of printer technology, but the innovation of printing materials. Currently, printing materials are mainly based on memory alloy material. The emergence of 4D printing technology will inevitably push the development of the field in new material technology. The recent research results of 4D printing technology, which was the first new technological achievement, were collaborated by MIT and Stratasys' Education & Research & Development. It was a revolutionary new technology that allowed materials to self-transform and self-assemble without the help of external forces. This technology in China is currently non-existent.

According to the founder of 4D printing technology, Tibbits and his team, the current 4D printing is still in its infancy. It can only print strand-like objects to realize the changes from one dimension to two dimensions and from one dimension to three dimensions. Moving forward, they will design and print out sheet-like objects and then three-dimensional objects to realize changes from two dimensions to three dimensions and three dimensions to more complex structures. Hence, the Transformers that exist in science fiction films will come out of the screen and enter our lives.

It is conceivable that in the era of 4D printing, products will become more intelligent and humane. Space stations and satellites will be able to realize self-assembly and self-repair, projects in dangerous areas will no longer require human participation, and everything, such as bridges, dams, highways, and houses, will be built on their own according to their design. People will only need to sit at the computer, designing products that suit their ideas and needs, and then simply click "print," and then it's all going to be done.

SECTION 9 | OPPORTUNITIES AND CHALLENGES OF "WORLD FACTORY"

The significance of industrial manufacturing to human society is self-evident. It is clear that industry possesses immense creativity and penetrates nearly every domain, industrializing various aspects of modern human life. From agriculture, forestry, fisheries, and transportation to information transmission, culture, and arts; from education, healthcare, sports, and fitness to leisure and tourism, and even military warfare, industrialism is pervasive and dependent on industrial technology.

The greatest contribution of industry to humanity is that it serves as the vehicle and essential tool for technological innovation. Humanity's greatest scientific discoveries, technological inventions, and exceptional imagination require industry as their foundation and means. Technological advancement is the soul of industry, and industry is the body of technological progress, with the majority of technological innovations manifesting as industrial development or requiring industrial development as a prerequisite.

Compared to agriculture, which is limited by relatively fixed outputs, and commerce, which must be based on industry, industry naturally becomes the sector with the most potent regenerative capability among the three industries, playing a crucial role in the sustained prosperity of the economy and social stability. In fact, the reason why the bourgeoisie was able to create more productive forces in less than a century of class rule than all previous generations combined is due to the rapid development of industrial productive forces in capitalist society.

Industrial development empowers humanity to transform nature and access resources more effectively, with its products being directly or indirectly used for consumption, significantly improving people's living standards. Thus, only industrialized countries can become innovative nations, and having a developed industry, especially manufacturing, is essential for leading technological innovation. The era of industrialization, marked by scientific rationality and technological progress, has been the most glorious phase of human development. It can be said that since the development of industrial civilization, industry has, in a sense, determined human survival and development. Given this and today's context, the competition among industrialized nations becomes especially significant.

"World Factory" is a specific term for the world's industrial powers, especially the manufacturing powers. In the history of the world economy, the three countries explicitly known as the "world factories" are the UK, the US, and Japan, in that order.

In 2009, China replaced Japan as the world's second-largest trading country after the US and successfully surpassed Germany to become the world's largest exporter. At the same time, China's manufacturing industry accounted for 15.6% of the global manufacturing value, becoming the world's second-largest industrial manufacturing country after the US. For a time, China became globally recognized as a "world factory."

However, times have changed. Nowadays, the cost of hiring a worker in China is currently 1.5 times higher than in Thailand, 2.5 times more than in the Philippines, and 3.5 times more than in Indonesia. China, as the "world factory," has the advantage of cheap labor costs that have gone up, and Southeast Asian countries are ambitious to become the "substitutes." Facing nearly three decades of smooth sailing and unmatched competitiveness, China's manufacturing sector now encounters a pincer challenge of high-end technology and low-end cost competition. With the traditional advantage of cheap labor no longer as prominent, China can rely on its extensive industrial infrastructure, a robust and evolving technological innovation system, and a large domestic market to continue maintaining its status as the "world's factory."

Meanwhile, developed countries like the US are loudly advocating for "reindustrialization" and "revitalizing manufacturing." Specifically for the US, which began its "deindustrialization" journey after World War II, as a traditional industrialized country that had completed its industrialization process before the war and started entering the post-industrial phase, the US changed its strategy from directly exporting electromechanical and automotive products to Western Europe to making substantial direct investments in Europe for localized production to circumvent the European Economic Community's tariff barriers.

The process of industrial hollowing out in post-war US actually reflects the deep trend of "moving from substance to form" in the country's industrial structure. During this process, the manufacturing sector continuously shrank and was considered a "sunset industry" in the US. Looking at the proportion of manufacturing in the national economy, there was a noticeable decline in the US manufacturing sector post-war. Apart from a few sectors like electronics manufacturing, traditional manufacturing sectors such as machinery manufacturing and automotive manufacturing experienced a long-term decline in their proportion of the national economy. The virtual economy, which was supposed to serve the real economy, inflated uncontrollably.

Although the "deindustrialization" measures taken by Western countries led by the US were once considered wise, seen as an inevitable change when a country reaches the middle and late stages of industrialization with sufficient technological and capital accumulation and a high level of consumer spending, the harm of "deindustrialization" is now evident. Objectively, reindustrialization and the revitalization of manufacturing require policies that promote domestic manufacturing development, reduce the cost advantages of foreign competitors, and make manufacturing more attractive in the US.

In this context, the emergence of 3D printing and the arrival of 4D printing presents a glimmer of hope both for China, aiming to retain its "world factory" status, and for Western countries with "reindustrialization" needs.

The emergence of 3D technology will cause a rapid reduction in social manufacturing demand for industrial workers. Manpower will no longer be a scarce resource, and traditional factories will no longer exist. As long as there is a printer, production can be done anywhere. As a result, the added value of intellectual costs will further increase in the proportion of social manufacturing. Therefore, the US government is facing the emergence of manufacturing depressions and employment decline, which is particularly interesting in the new technological revolution represented by 3D and 4D printing.

Obama once proposed to promote a National Network for Manufacturing Innovation of as many as 15 manufacturing innovation institutions across the country. Each of these institutions had a research focus. One of the research focuses on the improvement of the relevant standards, materials, and equipment of 3D printing technology (also called additive manufacturing) to realize the use of digital designs for low-cost and small-batch production.

On August 16, 2012, the US government announced the establishment of a manufacturing innovation research institution jointly funded by the government and the private sector in Ohio. It mainly developed 3D printing technology to boost innovation and growth in the manufacturing industry and regained the leadership of the world industry. As the leader of the global economy, the US places its hopes on regaining the glory of the manufacturing industry through 3D printing, and the strategic significance of 3D printing is self-evident.

On March 24, 2014, Obama visited the 3D printed Canal House project during the 53-nation leaders' Nuclear Security Summit in the Netherlands. The project was already an attraction in the Netherlands, and it was no longer printing a pure model, but a full-size "model." These houses were initially tall and narrow brick houses built by wealthy Dutch traders along the canals in Amsterdam. But now, high technology makes it easy to restore them.

Similarly, the emergence of 3D printing and 4D printing gives hope but also poses a challenge for China.

In 2012, Vivek Wadhwa, Director of the Duke Center for Entrepreneurship and Research Commercialization in the US, published an article titled "The Future of Manufacturing is in the US, Not China" on the website of *Foreign Affairs* magazine, believing that "Technical advances will soon lead to the same hollowing out of China's manufacturing industry that they have to US industry over the past two decades." According to Vivek Wadhwa's point of view, the future economic model of humankind will be a creator economy with a 3D printing-based production model. There will no longer be large-scale industrialized production.

At this point, if China can seize the opportunity of 3D printing, its manufacturing industry will achieve more remarkable development. Conversely, if the opportunity is missed, it will not be long before the center of the manufacturing industry will return to the US, and China will further widen the gap between the Western countries.

In the high-tech field of 3D printing, God has given the Chinese nation a rare opportunity to compete equally and appear on the same stage as the US and Germany. This is because there is no precedent in this field. Both large and small enterprises start at the same starting line. Unlike other fields, Chinese enterprises are trapped by the "patent pool" set up by those large international enterprises.

"3D printing is one of the signs of the third Industrial Revolution, and it is also the first time in the history of our country that we have the opportunity to participate in the rise of a new Industrial Revolution." Dai Rong, the Academician of Shanghai Ninth People's Hospital affiliated to Shanghai Jiao

Tong University School of Medicine, expressed that the difference of 3D printing technology between China and developed Western countries was not very big. Western countries have developed multi-nozzle printers, and China can also do the same. Western countries have relevant patents, and China has many patents, as well. There are many printing materials that are even superior to those of Western countries.

The Global and China 3D Printer Industry Market Status Research and Development Prospect Analysis Report for 2023–2028 highlights that the US accounts for 40.4% of the global 3D printing industry scale, with Germany following closely behind, and China ranking third. As a latecomer to the 3D printing scene, China has, in recent years, focused on independent innovation and research and development. Although there is still a certain gap compared to international technology standards, China is gradually moving toward more refined and specialized development. The significant market potential within the country has also attracted the attention and investment of several foreign 3D printing industry giants, further promoting the development of China's 3D printing industry.

Although China and Western countries are on the same starting line of the emerging Industrial Revolution, it is undeniable that China lacks the environment for the growth of innovative talent, and the entrepreneurs lack the spirit of "persistence, professionalism, and dedication." So, it is a great concern for China to value the focus and investment in 3D and 4D printing technology.

First, the lack of innovation will restrict the demand for 3D printing in China. Processing manufacturing is still the mainstay of China's manufacturing industry, and product creativity and innovation are not accomplished overnight. There is also a contradiction between large-scale manufacturing and technological innovation. Second, China's 3D printing equipment lags far behind the international levels in terms of printing accuracy and equipment reliability, and the core components that restrict the development of 3D printing still rely on imports. Besides, the industry will have to encounter risks such as technological bottlenecks and investment withdrawal while maturing.

So now, China has to grasp opportunities and courageously face its challenges.

China is vigorously promoting the implementation of industrial transformation and upgrade strategies. Industrial transformation and upgrades are essentially an upgrade to the high-value link of the industrial chains, which means that China's industry must shift the advantage of competition from being low-cost driven to becoming innovation-driven. To complete this industrial upgrade, it must be driven and supported by three major innovations. The first is institutional innovation, which is the decisive condition for China to become an industrial power. Second is scientific and technological innovation, which is the key factor for China's industry to surpass the Western countries. The third is education innovation, which will serve as a guarantee for China's industry to have a large number of innovative talents to build industrial power.

This world has been quietly changing. If you find this out one day, it means the world is already different. It's just like investing. In the investment world, while looking back, you will see so many investment opportunities around you, and you will start to regret why you haven't chosen that investment opportunity. But looking forward, it's all abyss. Any business is full of risks, and opportunities are only reserved for those with judgment and foresight.

AFTERWARDS

When I first published *4D Printing: Changing the Future Business Ecosystem* in 2015, it was the world's first book on 4D printing technology. Eight years have passed since then, yet many people are still not fully aware of this significant, disruptive technological change occurring around us. Compared to the widespread adoption of 3D printing, the emergence of 4D printing might seem quieter, but it opens a door to a much broader realm of imagination. It will lead the third industrial revolution initiated by 3D printing, and the incorporation of the "time" dimension in 4D printing will enable a true transformation of the manufacturing industry. If quantum technology is a disruptive theory and technology to classical physics, then 4D printing represents a disruptive theory and technology to our traditional industry.

In this new edition, I define the book as *The Future of 4D Printing: Innovations and Applications*, illustrating how we are living in an age where science fiction becomes reality. With a divergent thinking approach, this book combines theory and practice; from what 4D printing is to how it can be done, the impact of 4D printing on modern medicine, military, design, manufacturing, energy utilization, and more, and the changes 4D printing will bring to human society. It also addresses legal and intellectual property issues related to 4D printing, as well as the changes 4D printing will bring to China and earth at large. Systematically presenting a different human society supported by 4D printing technology.

Once the first book on 4D printing in China and a pioneering work in the global field of 4D printing, this new edition focuses more on the current applications of 4D printing technology and the changes it brings. The book introduces domestic and international research, applications, and thought processes on 4D printing technology into our perspective. The envisioned future society under 4D printing is full of endless possibilities.

REFERENCES

Bodaghi, M., Damanpack A., and Liao W. "Self-expanding/shrinking Structures by 4D Printing." *Smart Materials and Structures* 25, no. 10 (2016): 105034.

Bodaghi, M., Damanpack A. R., and Liao W. H. "Adaptive Metamaterials by Functionally Graded 4D Printing." *Materials Design* 135 (2017): 26–36.

Chen, Gen. *4D Printing: Changing the Future Business Ecosystem.* Beijing: Machinery Industry Press, 2015.

Chen, Hualing, Luo Bin, Zhu Zicai, et al. "Research Progress in Intelligent Materials and Structural Additive Manufacturing in 4D Printing." *Journal of Xi'an Jiaotong University* 52, no. 02 (2018): 1–12.

CITIC Securities. "In-Depth Research Report on the 3D Printing (Additive Manufacturing) Industry."

Dang, Jingyi, Zhu Dongze, Liu Xincheng, et al. "Medical Applications and Prospects of 4D Printing Technology." *PLA Medical Journal* 46, no. 6 (2021): 598–602.

Huang, Yuxiang, Li Qi, Ye Wu, et al. "Research and Application Progress of 4D Printing Technology in Cardiovascular Tissue Engineering." *Journal of Biological Engineering* 39, no. 10 (2023): 4046–4056. doi:10.13345/j.cjb.230240.

Li, Qinglian, Meng Shuai, Feng Gang, et al. "Research Progress on Memory Polymer Materials for 4D Printing." *Journal of Jiangxi Normal University (Natural Science Edition)* 44, no. 6 (2020): 561–566. doi:10.16357/j.cnki.issn1000-5862.2020.06.02.

Ma, Jiong, and Zhang Quanli. "4D Printing Will Trigger a New Military Revolution."

REFERENCES

MarketsandMarkets. "Materials, Industry, and Regional Trends and Forecasts for the 4D Printing Market, 2019–2025." *Foreign Affairs* 3, no. 2 (2015): 1–5.

Miao, S., Zhu W., Castro N. J., et al. "4D Printing Smart Biomedical Scaffolds with Novel Soybean Oil Epoxidized Acrylate." *Scientific Reports* 6 (2016): 27226.

Momeni, Faris, Hassani Navid S. M., Liu Xin, et al. "A Review of 4D Printing." *Materials Design* 122 (2017): 42–79.

Quadri, F., Soman S. S., and Vuyavenkataraman S. "Progress in Cardiovascular Bioprinting." *Artificial Organs* 45, no. 7 (2021): 652–664.

Rastogi, P., and Kandasubramanian B. "Breakthrough in the Printing Tactics for Stimuli-Responsive Materials: 4D Printing." *Chemical Engineering Journal* 366 (2019): 264–304.

Ren, Zhenghe, He Zhen, and Chen Shaojun. "Advances in the Application of Shape Memory Polymers in 4D Printing Technology." *Polymeric Materials Science and Engineering* (2024): 1–12. https://doi.org/10.16865/j.cnki.1000-7555.2023.0250.

Rui, Bochao, and Wang Rui. "The Future Development Path of 4D Printing Technology." *China High-Tech* (2022): 58–60.

Ruzan, Naif. "Preparing for the 4D Printing Revolution." *Foreign Affairs* 6, no. 2 (2014): 12–15.

Shen, Guoji, Wang Ping, et al. "Broad Prospects of Military Applications of 4D Printing Technology." *Science and Technology Daily* (December 5, 2016).

Tibbits, Skylar. "4D Printing: Multi-material Shape Change." *Architectural Design* 84, no. 1 (2014): 116–121.

———. "Printing Products." *3D Print Add. Manuf.* (2016): 135.

Truby, Ryan L., and Lewis Jennifer A. "Printing Soft Matter in Three Dimensions." *Nature* 540, no. 7633 (2016): 371–378.

Wang, Ruichen, Liu Xiujun, Zhang Jing, et al. "4D Printing of Stimulus-Responsive Shape Memory Materials and Its Research Progress." *Functional Materials* 52, no. 10 (2021): 10069–10074.

Xie, Lai. "4D Printing: Smarter Than 3D Printing." *International Pioneer Herald* (May 15, 2013).

Xu, Ke, and Li Xiaohong. "4D Printing Technology and Its Prospects in Military Applications."

Xu, Wang. *4D Printing: From Concept to Reality.*

Zhang, Yumeng, Li Jie, Xia Jinjun, et al. "4D Printing Technology: Processes, Materials, and Applications." *Materials Review* 35, no. 1 (2021): 1212–1223.

Zhang, Jifeng, Yin Zhifu, Ren Luquan, Liu Qingping, Ren Lei, Yang Xue, and Zhou Xueli. "Advances in 4D Printed Shape Memory Polymers: From 3D Printing, Smart Excitation, and Response to Applications." *Advanced Materials Technologies* 7, no. 9 (2022).

REFERENCES

Zhao, Meng, Wang Yongxin, and Liang Jin. "Research Progress of 4D Printing Technology." *Metalworking (Hot Working)* (2020): 32–36.

Zhao, Wei, Yue Chengbin, Liu Liwu, Liu Yanju, and Leng Jinsong. "Research Progress of Shape Memory Polymer and 4D Printing in Biomedical Application." *Advanced Healthcare Materials* (2022).

Zhao, Yubo, Zhang Lang, Chen Qian, et al. "Research Progress of 4D Printing Technology in the Food Processing Field." *Food Science* 44, no. 5 (2023): 338–345.

INDEX